Food Microbiology

NIPA® GENX ELECTRONIC RESOURCES & SOLUTIONS P. LTD.
New Delhi-110 034

About the Author

Dr. N. Ramanathan, obtained his masters degree and Ph.D in Microbiology from Annamalai University and served as Professor and Head, Department of Microbiology, Annamalai University, Tamil Nadu (2012-2015). He has around 27 years of UG and PG Teaching and Research experience. He has published more than 40 research papers in reputed journals and guided 40 M.Phil and 9 Ph.D candidates. He has attended 50 International and National conferences and delivered special lectures. He has organized two National seminars on Applied Microbiology. He was conferred the 'Best teacher' award by the university. He is life member of Indian Association of Applied Microbiologists and as Principal investigator conducted UGC major research projects. He is the author of text books on Principles of Microbiology, Environmental Microbiology, Objective Agricultural Microbiology and Objective Medical Microbiology. He served as Chairman and member of various academic bodies in several reputed universities and colleges.

Food Microbiology

N. Ramanathan
Former Professor and Head
Department of Microbiology
Annamalai University
Annamalai Nagar-608 002
Tamil Nadu, India

NIPA® GENX ELECTRONIC RESOURCES & SOLUTIONS P. LTD.
New Delhi-110 034

NIPA® GENX ELECTRONIC
RESOURCES & SOLUTIONS P. LTD.

101,103, Vikas Surya Plaza, CU Block
L.S.C. Market, Pitam Pura, New Delhi-110 034
Ph : +91 11 27341616, 27341717, 27341718
E-mail: newindiapublishingagency@gmail.com
www: www.nipabooks.com

For customer assistance, please contact
Phone: + 91-11-27 34 17 17
Fax: + 91-11- 27 34 16 16
E-Mail: feedbacks@nipabooks.com

ISBN: 978-81-19103-71-3

Composed and Designed by NIPA®.

Preface

The biodiversity of foods-cereals, pulses, oilseeds, spices, fruits and vegetables is varied and high in India, where we have varied agro-climatic regions from Kashmir to Kanyakumari. India is one of the World's major food producers and total food production in India is likely to double in next ten years. Green revolution (crop production), White revolution (milk production) and Blue revolution (fish production) were responsible for our country to meet the national requirements in food production. Biotechnological application in Agriculture and Animal Husbandry are going to play a pivotal role in further enhancing the food production.

Man has exhibited much thought and foresight in cultivating a variety of grains, fruits, vegetables, nuts and oilseeds and in rearing birds and animals for use as food. Food provides a rich habitat for microorganisms, as most foods contain an abundance of nutrients such as carbohydrates, proteins and lipids, vitamins and minerals. Microbial association with foods can be of desirable or undesirable nature in relation to the food product. However, foods are prone to microbial contaminants, which are known for their undesirable changes in foods-food spoilage and food poisoning. Although chemical changes and microbial growth usually mean food spoilage, in some cases, fermentation is desirable and microorganisms are actually added to foods. For example, in the production of fermented foods such as wine, bread, yoghurt, fish sauce, pickles, sausages, soya sauce etc. Microorganisms themselves are used as food (SCP, mushrooms) and their chemical products (organic acids, enzymes) are used in food processing. The plant, animal and microbial foods are now genetically modified to improve their qualities.

Food microbiology is a branch of applied microbiology and the scope of food microbiology is expanding rapidly to protect food from microbial spoilage and provide safe, nutritious food to consumers. We now live in a period of world-wide food crisis, a food saved is a food produced.

The book embodies twenty chapters covering the types and sources of microorganisms in food, factors influencing microbial growth in foods; Preservation of food by high temperature, low temperature, dehydration, osmotic pressure, irradiation, high pressure processing, chemical preservatives, food storage and packaging; Food safety and quality management, fermented food products, dairy microbiology, microbial foods and chemicals, mushroom cultivation, and microbial enzymes; Contamination, preservation and spoilage of cereals, sugars, fruits and vegetables, meat, sea foods, eggs and poultry, milk and milk products, canned foods; Bacterial food poisoning, mycotoxin and impact of genetically modified foods with descriptive and objective questions. In addition, procedures for fifteen practical experiments in food and dairy microbiology and glossary are included.

The overall objective of this book on 'Food Microbiology' is to bring together information on different areas of food microbiology in a single source. This book would be of great help to under graduate and post graduate students in the disciplines of Microbiology, Agriculture, Food technology, Biology and Biotechnology.

N. Ramanathan

Contents

1

BASIC PRINCIPLES OF FOOD MICROBIOLOGY

1.1. IMPORTANCE OF FOOD MICROBIOLOGY

Food may be defined as substances that provide nutrients, which when eaten and absorbed by the body, maintain life and growth. Food is basically energy stored in plant and animal foods in chemical form. On consumption, the stored form gets converted to physiological energy. Food is the prime necessity of life. Man has exhibited much thought and foresight in cultivating a variety of grains, fruits, vegetables, nuts and oilseeds and in rearing birds and animals for use as food.

Of the three primary needs of an individual, food, clothing and shelter, 'food' occupies the most important place. The biodiversity of foods- cereals, millets, pulses, oilseeds, spices, fruits and vegetables is varied and high in India, where we have varied agro-climatic regions from Kashmir to Kanyakumari. All the fifteen major climates in the world exist in India. India is the second largest producer of food in the world. The impact of green revolution is remarkable in India with many fold increased food production in the last five decades. The full potential of Indian agriculture can not be realized unless India emerges as a major player in the World agriculture.

World population is set to go beyond nine billion by 2050 and demand for food is likely to increase further because of growing affluence and globalization. Food security is achieved when all people have physical and economic access to sufficient, safe and nutritious value added food. India is emerging as a food power in the international trade of foods. The food processing industry in India has been identified as one of thrust areas for the development of the country. It has emerged as a highly perspective sector with immense potential in food processing. The future of Indian food processing industries is bright, provided it improves its quality of products as per international standards.The quality of food processing refers to the adherence of physical, chemical, biological, nutritional and sensory parameters required to achieve the wholesomeness of a food product. Quality control has become an important aspect for food processing industry for both domestic and export markets.

The understanding of biological principles of microflora associated with plants and animals in their natural habitat is of great significance since human food sources are basically of plant and animal origin. The knowledge about the precise beginning of human awareness of the presence and role of microorganisms in food was known before the establishment of microbiology as a science. Foods occur in diverse forms comprising mainly of *cereals and cereal products, sugar and sugar products, meat and meat products, fish and sea foods, eggs and poultry, milk and milk products, fruits and vegetables, canned foods and bottled beverages.* Microorganisms are ubiquitous in nature and found in human food also. Food is a suitable medium for growth and activity of variety of microorganisms. Food provides a rich habitat for microorganisms, as most foods contain an abundance of nutrients such as carbohydrates, proteins and lipids and growth factors such as vitamins and minerals. Microbial association with foods can be of desirable or undesirable nature in relation to the food product. However, foods are prone to microbial contaminants, which are known for their undesirable changes in foods. These microbial contaminants are of spoilage and/or pathogenic types. In most cases microorganisms use our food supply as a source of nutrients for their own growth. By increasing their numbers, utilizing the nutrients, producing enzymatic changes, contributing off flavours, they can spoil a food.

The microbial activity in food may be beneficial or harmful. Four aspects of microbial activity are of relevance to processing and preservation of food.

i) **FERMENTED FOODS**: Microorganisms can be used as processing aids in the production of fermented foods. Using specific microorganisms under controlled conditions fermented foods with better shelf-life, stability, palatability, flavour and organoleptic properties are produced (e.g. bread, wine, cheese, vinegar, pickle)

ii) **FOOD CHEMICALS FROM MICROORGANISMS**: A variety of food chemicals and additives may be produced by fermentation (e.g. citric acid, acetic acid, lactic acid, amylase, protese, pectinase, microbial rennet, xanthan-polysaccharide, β-carotene-food colourant, etc.). In addition, microorganisms themselves may be used as food, mushrooms, Spirulina - single cell proteins (SCP), single cell oil (SCO).

iii) **FOOD POISONING AND FOOD INFECTIONS**: Pathogenic microorganisms grow in food and produce toxins (botulin, aflatoxin) which are detrimental to consumers. Food also serves as a medium for certain pathogens that cause food infections and diseases (salmonellosis).

iv) **FOOD SPOILAGE**: The growth and activity of microorganisms in food cause the spoilage of food through undesirable enzymatic changes affecting the quality of the food. The enzymatic changes include the formation of products which contribute off-flavours and affect the textural and keeping quality of the food.

Food spoilage may be defined as any change that renders food unfit for human consumption. These changes may be caused by various factors, including contamination by microorganisms, infestation by insects, or degradation by endogenous enzymes (those present naturally in the food). In addition, physical and chemical changes, such as the tearing of plant or animal tissues or the oxidation of certain constituents of food, may promote food spoilage. The typical microorganisms that cause food spoilage are bacteria (*e.g., Erwinia, Pseudomonas, Bacillus*), yeasts (*e.g., Candida, Zygosaccharomyces*), and molds (*e.g., Rhizopus, Aspergillus, Penicillium*). Microbial spoilage of food is a normal consequence of microbial activity. By this activity in nature, they can render our food supply unfit for consumption. To prevent this we minimize the contact between microorganisms and our foods (prevent contamination) and also eliminate microorganisms from our foods or at least adjust conditions of storage to prevent their growth (preservation).

In the case of contamination of human pathogens, their association with our food supply is critical from a public health point of view. Many of our foods will support the growth of pathogenic microorganisms or serve as vector of them. We take steps to prevent their entrance and growth in our foods or eliminate them by processing. It is presumed that right from the beginning, the problems of spoilage and food poisoning were encountered. In order to overcome these problems as well as increase the shelf life of foods, a majority of the foods are either canned or dehydrated or frozen or fermented or irradiated or chemically preserved.

In recent years, the consumer awareness has played a predominant role in emphasizing the need for microbiologically safe foods for human consumption. Serious health hazards due to the presence of pathogenic microbes in foods can lead to food poisoning outbreaks. All these contributing factors make food microbiology, a many faceted and a challenging area of scientific development. A food microbiologist studies the role of microorganisms in i) the spoilage of foods and food borne diseases and how these may be prevented ii) the manufacture of fermented foods, microbial foods and food chemicals and how to improve their quality. Food is not a question of protein, fat or calories. Food is about culture. Food is about heritage. Food is about health. And in this context, food needs to receive a different approach by Governments.

1.2. TYPES OF MICROORGANISMS IN FOOD

FOOD MICROFLORA The microorganisms which are of importance in food microbiology include bacteria, yeasts, molds, protozoa and viruses.

BACTERIA : Bacteria are most abundant organisms found in naturural and extreme habitats. They are prokaryotic unicellular organisms of aerobic or anaerobic in nature and exhibit many morphological forms. The three principal shapes of bacteria are spherical shape of cocci, rod shape of bacilli and spiral shape of spirilla and vibrios. All food bacteria are small in size, typically few microns in length or diameter. Few genera of bacteria form dormant endospores which are resistant to heat, presence of inhibitory chemicals and other adverse conditions during food processing. Most bacteria multiply best at ambient temperature (16-38°C) and are termed mesophilic. Psychrophilic bacteria can grow at low temperatures (0 to 15°C), while thermophlic ones can grow at higher temperatures (55 to 85°C). The following bacteria play significant role in food and grouped on the basis of their activity.

LACTIC ACID BACTERIA : Lactic acid bacteria (LAB) are constituted of a heterogenous group of bacteria, ferment sugars to lactic acid and include the species of *Leuconostoc, Lactobacillus, Streptococcus* and *Pediococcus.* They are normally present in milk, and are also used as starter cultures in the production of cultured dairy products. Their activity is desirable in a variety of foods such as sauerkraut, and dairy products for the production of flavour. They cause spoilage of wines. Lactic acid bacteria consist of a number of bacterial genera: *Carnobacterium, Enterococcus., Lactobacillus, Lactococcus, Lactosphera, Leuconostoc, Melissococcus, Oenococcus, Pediococcus, Streptococcus, Tetragenococcus, Vogococcus, Weissella.*

Members of LAB are Gram positive, non-sporulating, catalase negative, devoid of cytochromes but aero-tolerant, fastidious, acid tolerant and ferment carbohydrates into lactic acid and energy.Lactic acid bacteria are divided in to two groups, homolactic acid bacteria and heterolactic acid bacteria based on their metabolism and end products (Table 1.1). Homo-fermentative bacteria such as *Lactococcus* and *Streptococcus* follow glycolytic pathway to yield two lactates from one glucose molecule, whereas, heterofermentative bacteria - *Leuconostoc* and *Weissella,* follow 6 phosphogluconate pathway to transform a glucose molecule into lactate, ethanol, and carbon dioxide. Common homolactic bacteria include *Lactobacillus delbrueckii, Lactobacillus plantarum, Streptococcus faecalis, Pediococcus cerevisiae, Lactobacillus lactis,* and heterolactic bacteria include *Leuconostoc mesenteroides, Lactobacillus fermentum, Lactobacillus cremoris,*and *Lactobacillus brevis.*

TABLE 1.1. TYPES OF LACTIC ACID BACTERIA

TYPE	GENUS	MORPHOLOGY	GC CONTENT
HOMOFERMENTER	*Lactococcus*	Cocci in chains	33 - 37 %
	Pediococcus	Cocci	36 - 42 %
	Streptococcus	Cocci in chains	40 %
HETEROFERMENTER	*Leuconostoc*	Cocci	38 – 41 %
HOMO/HETEROFERMENTER	*Lactobacillus*	Rods	32 – 53 %

ACID BACTERIA: Bacteria generally produce acids. Vinegar (acetic acid) is the direct result of bacterial metabolism. Acetic acid bacteria need oxygen to convert the alcohol to acetic acid. Acetic acid bacteria, *Acetobacter aceti, Glucanobacter sp.* oxidize ethanol to acetic acid, useful in vinegar manufacture but are undesirable in alcoholic beverages.

Butyric acid bacteria, which produce butyric acid by fermentation of sugar, belong to the genus *Clostridium*. Propionic acid bacteria belong to the genus *Propionibacterium,* produce propionic acid by fermentation.

PROTEOLYTIC BACTERIA: Proteolytic bacteria include a heterogenous group of bacteria, which produce extracellular proteases. Most species belong to the genera, *Clostridium, Bacillus, Pseudomonas* and *Proteus. Bacillus cereus,* , *Pseudomonas fluorescens* are proteolytic bacteria , the former is aerobic spore former and the latter non-spore forming facultative aerobe.

LIPOLYTIC BACTERIA: Lipolytic bacteria are also a heterogenous group of bacteria, which produce lipases. Many of the aerobic proteolytic bacteria are also lipolytic. Most species belong to the genera, *Pseudomonas, Alcaligenes, Staphylococcus* , *Serratia* and *Micrococcus.*

SACCHAROLYTIC BACTERIA : They hydrolyze disaccharides and polysaccharides to simple sugars. *Clostridium butyricum* and *Bacillus subtilis* are examples of saccharolytic bacteria, which are also amylolytic, produce amylase to hydrolyze starch.

PECTINOLYTIC BACTERIA : These bacteria produce pectinases responsible for softening the plant tissue in fruits or loss of gelling power in various plant foods. Species belong to the genera, *Bacillus, Achromobacter, Aeromonas, Arthrobacter, Flavobacterium* are pectinolytic.

THERMOPHILES : Thermophilic and thermoduric bacteria are resistant to high temperature. The thermophilic bacteria cause spoilage of low acid canned foods. e.g. *Bacillus stearothermophilus* and *Clostridium thermosaccharolyticum.* Thermoduric bacteria survive heat treatment such as pasteurization. e.g. *Micrococcus, Staphylococcus, Microbacterium.*

PSYCHROPHILES : Psychrotrophic organisms have particular significance as pathogens or spoilage organisms as they can multiply during refrigeration or improper processing. Temperature abuse and incorrect time/temperature relations are main causes for microorganisms being not destroyed at the expected rate or even getting a chance of multiplying. Psychrotrophic and psychrophilic bacteria are able to survive and grow at refrigeration temperature. This means that keeping qualities or shelf lives of food depend greatly on the contaminated psychrotrophic organisms, as they can grow relatively rapid at low temperature ranges around 0°C. They spoil non-sterile refrigerated foods such as meat, poultry, egg and other foods. e.g. *Pseudomonas, Achromobacter, Alcaligenes, Chromobacterium* and *Flavobacterium.*

HALOPHILES : They are salt loving bacteria, include species the genera *Halobacterium, Halococcus, Haloarcula, Halobaculum, Bacillus, Vibrio, Moraxella, Corynebacterium, Streptococcus* and *Clostridium.* The absolute requirement of high concentrations of NaCl is the most striking feature of the halobacetria .They grow best at concentrations of 3.5 to 4.5 Molar NaCl. They are not only salt tolerant but also salt requiring. They require certain minimal concentrations of dissolved sodium chloride for their growth and survive at higher salt concentrations to different extents depending on their halotolerant ability. Osmophilc or saccharophilc bacteria grow in high concentrations of sugars. e.g. *Leuconostoc.* Halophilic bacteria have many adaptations to survive in the extreme environment. Salt tolerant membranes are stabilized by combinations of polar lipids, neutral lipids, divalent cations (Mg, Ca) and proteins. They have salt tolerant enzymes that exhibit strong hydrophobic interaction, which contribute to their stability. Structural integrity of cell is due to high internal concentration of salt, exclude Na^+ from interior and concentrate K^+. KCl is required for stability of enzymes. The members of alkalophilic bacteria (*Natronorubrum, Natronomonas*) grow at pH around 10.

PIGMENTED BACTERIA : Pigmented bacteria produce colour during their growth in foods. e.g. *Flavobacterium* (yellow to orange), *Serratia* (red), *Halococcus* (pink, red and orange). *Lactobacillus plantarum* produces rust colour pigment discolouring cheese, *Flavobacterium* sp cause the discolouration of meat, spoilage of shell fish, poultry, egg, butter and milk. *Pseudomonas fluorescens* causes green rot of eggs.

SLIME (ROPINESS) FORMING BACTERIA : These bacteria include *Alcaligenes viscolactis. Enterobacter aerogenes* and *Klebsiella oxytocalis* and some species of *Streptococcus* and *Lactobacillus,* which cause ropiness in milk. *Lactobacillus plantarum* causes ropiness in fruit and vegetable products, *Bacillus* cause sliminess in meat while *Gluconobacter oxydans* cause ropiness in beer.

GAS FORMING BACTERIA: Gas forming bacteria include species of the genera, *Leuconostoc, Lactobacillus and Propionibacterium* which produce carbon dioxide. Species of *Escherichia, Enterobacter, Proteus, Bacillus, and Clostridium* produce both carbon dioxide and hydrogen. *Clostridium spp* are involved in gaseous spoilage of canned vegetables and putrefaction of foods.

OFF FLAVOURS FORMING BACTERIA: *Streptomyces* spp produce undesirable flavour, musty odour and taste in foods and *Pseudomonas spp* produce a variety of metabolites that affect the flavour of foods dexterously.

COLIFORMS: They are group of intestinal bacteria that are aerobic or facultative anaerobic, Gram negative, non-spore forming rods, ferment lactose with gas formation. Coliforms are indicator organisms, closely associated with the presence of pathogenic organisms but not necessarily pathogenic themselves. *Escherichia coli* is a bacterium found in lower intestine of warm blooded animals. Most *E.coli* strains are harmless, but some, such as serotype 0157:H7, can cause serious food poisoning in humans. The harmless strains are part of normal flora of intestine (colon), and can benefit their hosts by producing vitamins or by preventing establishment of pathogenic bacteria in gut. *E.coli* is not always confined to the gut, and its ability to survive for brief periods outside the body makes it an ideal indicator organism. *E.coli* counts in foods are used as indicators of faecal pollution. They cause spoilage of variety of foods producing off-flavours and sliminess. e.g. *Escherichia coli* and *Aerobacter aerogenes.*

It is almost impossible to isolate from water the organisms responsible for waterborne diseases. Few are present and they donot multiply in water. The only safe method to prevent waterborne diseases is to condemn fecally polluted water as being unfit for human use, as it may contain harmful organisms. Fecal pollution can be determined by examination of water for colon bacilli (*E.coli*), abundant in feces and not found outside intestinal tract in nature. The colon bacilli (*E.coli*) in water indicate the presence of pathogenic microbes in water, which may be responsible for a number of waterborne diseases. Hence *E.coli* is known as **indicator organism**. Water also contain certain bacteria that resemble colon bacilli but may or may not be of fecal oirgin. These bacteria also ferment lactose with formation of gas like *E.coli*. Such bacteria are collectively known as a group, the **coliform bacteria**.

DISEASE CAUSING BACTERIA / FOOD PATHOGENS

A more serious aspect of microbial contamination is the possibility of serious health hazards due to potential pathogenic organisms gaining entry into foods due to low standards of hygiene and sanitation.

Clostridium botulinum is an anaerobic, Gram positive, endospore forming bacillus. It is a common inhabitant of soil and water, the usual sources from which foods are contaminated. **Botulism,** a very serious type of food intoxication disease is caused by *Clostridium botulinum. Staphylococcus aureus* is the causative organism of Staphylococcus enterotoxicosis, by the action of an enterotoxin. The toxin liberated is termed as enterotoxin because it causes gastroenteritis or inflammation of the lining of intestinal tract. *Bacillus cereus,* Gram positive, aerobic, spore forming rod, causes gastroenteritis food intoxication through the enterotoxin produced in contaminated food.

Salmonella spp are Gram negative, non-spore forming rods, motile, enteric. The most frequent species associated with food infection in humans is *Salmonella typhimurium.* Salmonellosis may result following the ingestion of viable cells of *Salmonella* spp. *S. choleruesuis* and *S. enteritidis* also cause salmonellosis.

Clostridium perfringens, Gram positive, nonmotile, anaerobic spore forming rod, causes food infection and the symptoms include acute abdominal pain, diarrhea, and gas. The enterotoxin produced by the bacterium during sporulation is heat sensitive. Spores are found in raw food, soil, sewage and animal faeces.

Vibrio parahaemolyticus, Gram negative, curved or straight motile rod, halophilic bacterium (requires 1-3 % NaCl), causes the food infection, Vibriosis. It is responsible for 60% cases of summer diarrhea in Japan.

Campylobacter jejuni, is a Gram negative, microaerophilic, thermophilic (42°C) spiral, sea gull shaped, highly motile cell that moves in cork screw fashion. It is the most common bacterial agent of Mucosanguineous diarrhea.

1. CYANOBACTERIA

Cyanobacteria are the simplest, prokaryotic microorganisms called as oxygenic phototrophic bacteria. They were previously grouped under algae and known as Blue green algae (BGA). Based on genetic and cytological characters, they are now grouped in bacterial classification. They occur principally in fresh water ecosystem and soil. Cyanobacteria grow well wherever water and sunlight are available. Cyanobacteria are found in varied and extreme environments such as artic ,antartic glaciers and tropical hot springs. Cyanobacteria are exploited as food (SCP), feed, fuel (hydrogen) and medicine. Cyanobacterium *Spirulina* is used as SCP- single cell protein.

2. YEASTS

Yeasts are single celled fungi (Fig 1.1). Over 600 different species of yeasts are known and they are widely distributed in nature. *Saccharomyces* spp are the most widely used yeasts in foods. *Saccharomyces cerevisiae* is used in the manufacture of many foods, with special strains are used for leavening of bread and for the production of ale, wine, alcohol, glycerol and invertase. *Saccharomyces cerevisiae* var *ellipsoideus* is a high alcohol yielding strain employed in the production of industrial alcohol, wine and distilled liquors. *Saccharomyces fragilis* and *Saccharomyces lactis* ferment lactose and used in the dairy industry. Yeast is probably the most widely used microorganisms for Single Cell Protein (SCP) production. These include strains of *Candida utilis, Candida arborea, Candida pulcherrima* and *Saccharomyces cerevisiae*. Yeasts are best known for their beneficial contributions to society, and the literature abounds with discussions of their role in the fermentation of alcoholic beverages, bread, and other products. Yeasts also cause spoilage, but, with a few exceptions, this unwanted activity often goes unrecognized and underestimated as a major problem in the food and beverage industries. *Rhodotorula spp*. cause discolouration of food by forming coloured spots on meat while *Candida lipolytica* spoils butter.

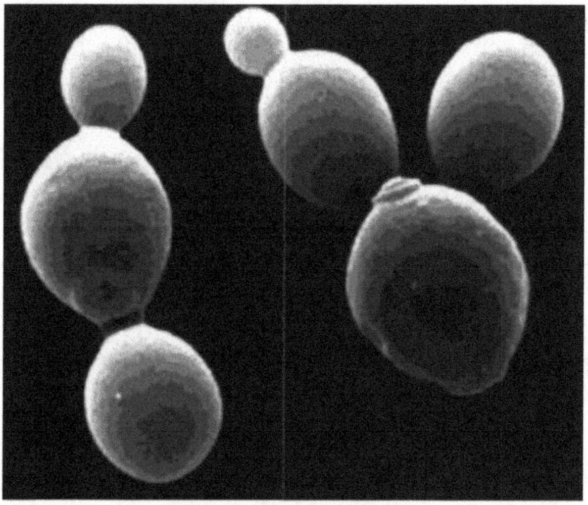

Fig. 1.1 Budding yeasts

MOLDS (FUNGI): Molds are eukaryotic, multicellular, mesophilic heterotrophic organisms, possess filamentous hyphae and network of mycelium. A large and widely distributed group of organisms, fungi perform activities essential to the functioning of all natural ecosystems.

They are among the foremost decomposers of organic matter, breaking down plant and animal remains and wastes into their chemical components. Fungi's value to humankind is inestimable. Certain types of fungi, including several types of mold, have proven extremely valuable in the synthesis of antibiotics and hormones used in medicine and of enzymes used in certain manufacturing processes. Some fungi, such as mushrooms and truffles, are considered tasty delicacies that enhance a wide variety of recipes. Not all fungi are beneficial-some damage agricultural crops, cause disease in animals and humans, and form poisonous toxins in food.

With the exception of one-celled species, most fungi are composed of threadlike tubular filaments called hyphae. Each individual hypha is surrounded by a fairly rigid wall usually made of chitin. Hyphae that are partitioned by dividing cross walls are called septate hyphae, and hyphae without cross walls are called nonseptate hyphae. Hyphae grow by elongation at the tips and by branching to form an interwoven mat known as the mycelium. As the mycelium develops, it may produce large fruiting bodies or other structures that contain reproductive spores. Fruiting bodies are often the most visible structure of a fungus, usually growing above the soil or other surfaces so that the spores can be dispersed by air currents or other mechanisms. In contrast, the mycelium is usually hidden beneath the surface of the plant, animal, food or other material it is decomposing. For example, a mushroom mycelium is typically buried beneath the soil surface, while its fruiting body, the familiar umbrella-shaped structure, sprouts from the ground.

Fungi have been used as a food source since the beginning of recorded history. Mushrooms add flavour, texture, and nutritional value to many dishes. Other fungi are used in the manufacture of foods. Certain molds are used to ripen cheeses, such as Brie, Camembert, and the characteristic blue-veined Roquefort. In Asia, fungi are added to soybeans and allowed to ferment to make several foods saccharification of starch and some species are used in the manufacture of oriental foods.

Aspergilli (Fig 1.2) are wide spread and involved in the spoilage of many kinds of foods. Mycotoxins are poisons produced by fungal growth in cereals, nuts, fruits, and vegetables. More than 100 species of fungi produce these toxins. The most common mycotoxin is aflatoxin, produced by *Aspergillus flavus* and *Aspergillus parasiticus*. Commonly found on corn, peanuts, and tree nuts, the toxin also can be transmitted to humans through the milk, meat, or eggs of animals fed contaminated grains. Aflatoxin is the most potent carcinogen, or potentially cancer-causing agent, yet discovered. Other mycotoxins include trichothecenes and zearalenone, compounds known to injure the intestines, bone marrow, lymph nodes, spleen, and thymus. They are produced by species of *Fusarium* that grow on grain, straw, or hay stored while damp. Occasionally, circumstances prevent the harvesting of grains during the autumn, and the grains lie dormant in the damp fields until they are harvested in the spring. These grains are especially vulnerable to trichothecenes and zearalenone contamination. A large outbreak of trichothecenes contamination occurred in Russia in early 1944 among hungry peasants who had been searching the winter fields for unharvested wheat and millet. Aflatoxins were first reported in Brazilian groundnut meal in early 1960s which when fed caused death of lakhs of Turkey birds in England.

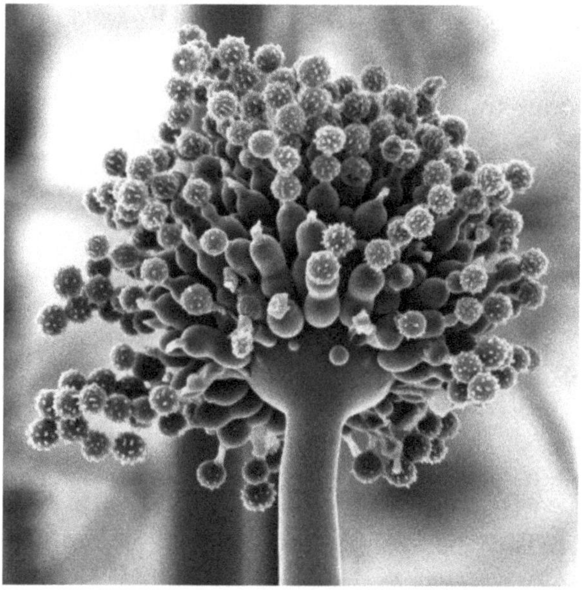

Fig. 1.2. Mold- Aspergillus

A.glaucus and A.repens, A.niger are used in the manufacture of citric and gluconic acids and many enzymes. Sake is the Japanese national drink prepared by fermenting rice with *A.oryzae* and the fungus is also useful in making oriental foods and production of enzymes.

Penicillium genus is wide spread in occurrence. Some species cause spoilage of foods while certain others are useful. *Pencillium oxalcium, P. islandicum, P. funiculosum* cause spoilage of cereals. *P. crustosum, P. commune, P.echinulatum* are found in lipid and protein rich foods -

Cheese, meat and nuts. *P. expansum, P.hirsutum, P. italicum* are found on fruits and vegetables. *P. camemberti* and *P. roquefortii* are useful in ripening of cheeses. Other commonly occurring spoilage fungi are: *Botrytis cinerea* (grey mold rot), *Alternaria* (black rot), *Trichoderma* (green mold rot), *Fusarium* sp. (brown rot), *Colletotrichum* sp. (anthracnose).

3. PROTOZOA

Protozoa are highly diverse group of unicellular, non-photosynthetic eukaryotic microorganisms. Majority of the protozoa are free living organisms found in moist habitats such as salt water, fresh water, soil and decaying organic substrates. Few species are parasitic , cause devasting human diseases like Malaria. Many protozoa form a dormant stage called **cyst** to withstand adverse conditions such as low nutrients, low pH, absence of oxygen. Amebiasis is a food and water borne disease caused by *Entamoeba histolytica.*

VIRUSES: Viruses are considered as acellular infectious agents, incapable of independent growth in artificial media. They live on other organisms like animals, plants, microbes and utilize the machinery and chemicals of invaded host cells for multiplication. Viruses contain a bit of nuclear material, either DNA or RNA, and never both. The nucleic acid is surrounded by a protein coat called CAPSID, made up of capsomeres. i) The capsid protects the genetic material, when the virus is outside the host cell. ii) It serves as vehicle for the entry in to another host. Hence viruses can also be called as 'Nucleocapsid'. The structurally complete, mature and infectious virus is called ' Virion". Some are enclosed by an envelope, some viruses have spikes. Most viruses infect only specific types of cells in one host. Host range of viruses is determined by specific host attachment sites and cellular factors. Viruses infect humans, animals, plants, fungi, protozoa, algae, and bacterial cells.Only few viruses are associated with food. Polio virus and hepatitis viruses A and E are water and food borne. Rota viruses and Norwalk viruses are water borne and cause viral gastroenteritis.

1.3. PRIMARY SOURCES OF MICROBES FOUND IN FOOD

Microbial contamination in foods

Bacteria and fungi (yeasts and molds) are the principal types of microorganisms that cause food spoilage and food-borne illnesses. Foods may be contaminated by microorganisms at any time during harvest, storage, processing, distribution, handling or preparation. Microbial association with foods can be desirable or undesirable nature in relation to the final product. However, foods are prone to microbial contaminants which are known more for the undesirable changes which they bring about in foods. These microbial contaminants are of spoilage and/or pathogenic types. The spoilage activities group of microorganisms manifest in the form of defects arising out of the action of microbial enzymes on food constituents giving rise to various off-flavours such as bitter, putrid, rancid and fruity. A more serious aspect of microbial contamination is the possibility of serious health hazards due to the presence of potential food poisoning microorganisms and their toxic metabolites leading to food poisoning out-breaks. The magnitude of microbial contamination is so large, that it is difficult to cover the entire group of microorganisms.

Many kinds of agents are potentially destructive to the agreeable or healthful characteristics of fresh foods. Microorganisms, such as bacteria and fungi, rapidly spoil food and their sources are many. In general there exists a close relationship between the source and the type of contaminant leading to the predominance in the given environment (source: sewage - organism *E. coli*).

The primary sources of microbial contamination are soil and water, plant and plant products, food processing machinery and utensils, intestinal tract of man and animals, food handlers, animal feeds and hides, air and dust.

1. Soil and water

Many fruits and vegetables are in contact with soil and become infected from that source. The underground parts like roots, tubers or bulbs as in radish, beet, carrot, potato, onion, ginger, garlic are in direct contact with moist soil and fruits like strawberries, cucumbers, tomatoes, melons are in direct contact with surface soil. Soil and water are two important habitats for microorganisms, which harbour millions of bacteria and fungi per gm or ml. Food microflora from soil and water are one and the same disseminated by wind from soil to water and water to soil. Food borne bacteria found in soil and water are

Achromobacter	*Acinetobacter*	*Alcaligenes*
Bacillus	*Clostridium*	*Corynebacterium*
Enterobacter	*Micrococcus*	*Pseudomonas*
Proteus	*Sarcina, Serratia,*	*Streptomyces*

Clostridium botulinum is a common inhabitant of soil, the usual source from which foods are contaminated. In general fungi are wide spread in soil. The common ones are: *Aspergillus flavus, Botrytis, Fusarium, Rhizopus, Penicillium, Trichothecium.* Yeasts are found in soil and their number is low in water.

Most of the water currently used comes from surface water supplies, which is most easily contaminated by human activity. From microbiological point of view, water quality studies would cover the hygiene aspect, which has its effects on public health. The sources of microbes in water are many, soil, air, excreta of animals, decaying plants and water itself. As a result, many different kinds of microbes are found in water- bacteria, fungi, protozoa and viruses, pathogens and as well as harmless water microflora like psychrotrophs. Water-borne pathogens are *E-coli* 0157-H7, *Vibrio cholerae, Salmonella typhi, Salmonella paratyphi, Salmonella enteritidis, Shigella dysentriae,* Polio virus, Hepatitis A virus, *Entamoeba histolytica* and *Ascaris lumbricoides.*

2. Plant and plant products

Soil and water constitute the primary sources of microorganisms to plants. Most of the soil microorganisms, listed above are also found on plants. Some bacteria are more associated with plants than with soil. They are *Acetobacter, Aerobacter, Erwinia, Flavobacterium, Lactobacillus, Leuconostoc, Pediococcus, Streptococcus, Xanthomonas.*

Among the molds, Soft rot disease causing fungi like *Fusarium, Mucor, Rhizopus* are associated with plants. Leaves, stems, flowers as in greens, lettuce, cabbage, asparagus, cauliflower are generally infected by plant pathogens. Plant pathogens are carried over from the field and cause post harvest spoilage of fruits and vegetables. Yeasts, *Saccharomyces, Rhodotorula, Torula* are found in fruits.

3. Food equipments and utensils

The types of microorganisms found on equipments and utensils depend on the type of food handled, care of the utensils and their storage. Cleaning them in hot water and store those free from dust in closed area are essential. The left-over food in the processing equipments and utensils attract contaminants and act as source of spoilage organisms in food.

4. Intestinal tract of man and animals

Microflora found in intestine of man and animals find their way on to plants, soil, water, utensils to food. The genera common to intestinal tract include *Escherichia, Enterobacter, Clostridium, Citrobacter* and *Pseudomonas.* Molds are not transmitted by faecal sources. Yeast *Candida* is common in intestine of man. *Salmonella* are abundant in the intestinal tract of poultry, a very important source and the *Salmonella* contaminated eggs may be a source of the organism in egg products.

5. Food handlers

The microflora of food handlers reflects the environment and habits of individuals. They are picked up from garments, soil, water, air and dust. The contaminant coming from food handlers is of serious concern as it comprises of mainly of pathogenic microorganisms. The pathogenic microbes are found in infections of body surfaces - hands and arms, nasal and throat discharges of sick food handlers. Typhoid fever, diarrhea, diphtheria are transmitted through milk from food handlers.*Micrococcus* and *Staphylococcus* are generally found specially associated with hands, nasal cavities and mouth. *Salmonella* and *Shigella* are intestinals deposited to food by food handlers if sanitary practices are not followed.

There are three sources of pathogenic bacteria causing human health hazard through food handlers i) Faecal matter and/or urine of infected humans, not only those showing symptoms, but also active asymptomatic carriers. ii) Nasal and throat discharges of sick individuals and asymptomatic carriers. iii) Infection on body surfaces of food handlers (hands and arms).

6. Animal feeds and hides

Animal feeds are of great importance in spread of food poisoning-aflatoxicosis and salmonellosis. The aflatoxin produced by *Aspergillus flavus* and *A .parasiticus* in animal feeds like groundnut cake is transmitted to human beings through milk. Organisms associated with feed are rapidly disseminated throughout processing plants. Bovine tuberculosis comes from consumption of milk of tuberculous cows. Organisms associated with soil, water, dust and faecal matter and feeds are found on the hides of animals. From hides they may get deposited onto hands of workers handling food and slaughtered animals. Bacteria most often involved in the spoilage of sea foods are part of the normal flora of the external slime. Bacteria of the genera, *Bacillus, Escherichia, Flavobacterium, Listeria, Micrococcus, Proteus, Serratia, Pseudomonas* and *Vibrio* are found in sea foods.

7. Air and dust

The contaminants coming from atmosphere are mainly the spore formers. Many bacteria and fungi are found in air and dust. Air receives the microorganisms from soil, water and vegetation. *Bacillus, Micrococcus, Streptococcus* and *Sarcina* species are commonly found in air, can endure dryness hence called dust bacteria.

The number and types of microorganisms present in a finished food product are influenced by the following i) Environment from which the food was originally obtained ii) Microbiological quality of the food in its raw or un-processed state iii) Conditions of hygiene and sanitation under which the product is processed and handled. iv) Adequacy of subsequent packaging, handling and storage conditions in keeping the microflora at a low level.

1.4. FACTORS INFLUENCING MICROBIAL GROWTH IN FOOD

Food as a substrate for microorganisms and the factors influencing food microflora

Natural food sources have all evolved mechanisms of defense against the invasion and proliferation of microorganisms and some of these remain in effect in fresh foods. One makes effective use of these factors in preventing the microbial spoilage of foods. The food or substrate's characteristics dictate what can grow or cannot grow. By understanding the characteristics of food/substrate we can make predictions about the microflora that may develop. The knowledge of the factors that favour or inhibit the growth of microorganism is essential to an understanding of the principles of food spoilage and preservation. The following chief compositional factors of food - intrinsic factors-influence the microbial activity: 1. pH, 2. Moisture, 3. O-R potential, 4. Nutrient content, 5. Inhibitory substances in foods, and 6. Biological structure.

1. pH

It has been well established that most microbes grow best at neutral pH [6.6-7.5]. Most spoilage bacteria grow best at neutral pH, pathogenic bacteria even more narrow in tolerance range at near neutral pH. Yeasts and molds have much greater tolerance to acidic pH. On either end of pH range organisms will experience increased lag and generation times.

Most foods are neutral in pH and readily spoiled by bacteria and molds. The meat with normal post rigor mortis pH of 5.8 have much larger shelf life than meat with a higher pH of 6.0 -6.5. Acidic foods like fruits, soft drinks, vinegar, wines, fermented milk and pickles are having good keeping quality. The acidity in fermented milk, pickles, vinegar is termed as **Biological acidity** as the acidity is due to the action (fermentation) of certain microbes. The acidity in fruits is due to organic acids of plant origin (citric acid in lime) and termed as **Natural acidity**. The bacteriostatic effect of a given pH depends on type of acid present acetic or lactic < citric or tartaric. Fruits have pH values below those required by most spoilage organisms, as a natural protection against spoilage. In foods with pH below 4.0, such as fruit juice and yoghurt, yeasts are likely to be dominant, although some lactic acid bacteria are equally successful in these conditions.

Based on the pH, the foods are classified in to three types, non-acid foods, low acid foods and high acid foods (Table 1.2).

TABLE 1.2. CLASSIFICATION OF FOOD BASED ON pH

S. No.	FOOD	pH	SPOILAGE MICROBES
1	**Non acid foods:** Eggs, Milk, Vegetables	>7.0	Bacteria, Molds
2	**Low acid foods:** Meat, Sea foods, Poultry	5.6- 7.0	Proteolytic bacteria, molds
3	**High acid foods:** Fruits, Cheese, Sauerkraut	< 5.6	Yeasts, molds, Lactic acid bacteria

pH affects the functioning of microbial enzymes and the transport of nutrients. Important cellular components such as DNA and ATP require neutrality. Few food spoilage fungi can thrive in wide pH range, from pH 2.0 to 9.5 and pH is not a restriction for their growth. e.g. *Penicillium funiculosum* can grow even at pH 1.8.

2. Moisture

Microbes require water for their growth and multiplication. The water contained in foods consists of bound water, which is chemically absorbed to the substrate and free water, which is weakly bound and more readily available for microbial growth.

Preservation by reducing the water content of food is one of the earliest examples of food biotechnology. It may be achieved directly by dehydration [sun or oven drying]. Although the terms, drying and dehydration are applied to the removal of water from food, to the food technologist drying refers to natural desiccation, such as spreading fruit on racks in the sun, and dehydration designates drying by artificial means, such as a blast of hot air. In freeze-drying a high vacuum is maintained in a special cabinet containing frozen food until most of the moisture has sublimed. Removal of water offers excellent protection against the most common causes of food spoilage. Microorganisms cannot grow in a water-free environment, their enzyme activity is inhibited and most biochemical reactions are greatly retarded.

Water requirements of microbes are expressed in terms of water activity, $\mathbf{a_W}$. The **water activity,** a_W is defined by the ratio of vapour pressure of water in a product (within food) to the vapour pressure of pure water at the same temperature. The a_W for pure water is 1.0, of most fresh foods is 0.99 and of saturated solution of NaCl is 0.75.The probability of microbial development is very high at $\mathbf{a_W}$ values near 1 and it decreases as the $\mathbf{a_W}$ decreases.

In many foods, water activity is the dominant factor governing the food stability and microbial spoilage of the food. At high a_W value of 0.95, bacteria are the dominant flora and foods with a_W values below 0.85 are rarely spoiled by common bacteria. The exceptions are salted sea foods that may be spoiled by halophilic (salt loving) bacteria (*Halobacterium halobium)* and salt loving fungi (*Polypaecilum pisce, Basipetospora halophila).* Gram negative bacteria are most sensitive to the decrease in water activity than Gram positive bacteria. At values below 0.90 , yeasts and molds take over the as the major spoilage organisms, particularly in sugar rich foods. Yeasts do not grow below 0.87 and molds not below 0.80. The minimal a_W values for spoilage organisms are given in the Table 1.3.

TABLE 1.3. MINIMAL a_W VALUES FOR SPOILAGE ORGANISMS

MICROBIAL GROUP	a_W	EXAMPLE	a_W
Many bacteria	0.95	*Clostridium botulinum*	0.97
Many yeasts	0.88	*Candida scottii*	0.92
Molds	0.80	*Aspergillus flavus*	0.80
Halophilic bacteria	0.75	*Halobacterium* sp	0.75
Xerophilic fungi	0.65	*Xeromyces bisporus*	0.65
Osmophilic yeasts	0.60	*Zygosaccharomyces* sp	0.62
No Microbial Growth	0.50		

Osmophilic yeasts have high tolerance for high solute concentrations. They concentrate polyols as osmoregulators and enzyme protectors. Osmophilic yeasts, *Zygosaccharomyces rouxii, Z. bisporus, Z. baili* cause spoilage of sugar syrups, fruit concentrates and chocolates. *Zygosaccharomyces rouxii* spoil soft filled chocolates; produces gas in the filling, splitting the casing and allowing leakage. *Zygosaccharomyces rouxii* can tolerate relatively high levels of SO_2 used in fruit products to prevent mold growth. The osmophilic yeasts found growing in media with an a_W as low as 0.62 .For enumeration of osmophilic yeasts, Scarr's osmophilic agar, MY50G medium, PDA with 60% sucrose medium are generally used.

Food preservation is achieved by the addition of solutes like sugar or salt. Since ancient times table salt has been a preservative for fish, ham, and bacon, and sugar has been used to preserve jelly, fruit jams, and fruit preserves. In solutions having high osmotic pressure water will leave the cell and cells get plasmolysed. This plasmolysis principle is practically applied in the preservation of jam, jellies, condensed milk by adding sugar and in meat, beef, sea foods and pickles by the use of salt. The water activity of different foods are given in the Table 1.4.

TABLE 1.4 WATER ACTIVITY OF DIFFERENT FOODS

Dates with 24% moisture	0.60
Raisins with 17% moisture	0.68
Honey	0.70
Fruit cakes	0.75
Jam and jellies	0.75
Hams and cured meat	0.80
Orange juice, concentrates	0.80
Cheese	0.89
Milk	0.97
Distilled water	1.00

3. O-R potential (Oxidation Reduction Potential)

OR potential of a substrate may be defined as the ease with which the substrate gains or looses electrons. When it looses electrons, it is oxidized and while it gains electrons it is said to be reduced. The amount of oxygen and carbon dioxide in the atmosphere, especially of dissolved oxygen in the substrate strongly influence growth of microbes.

The OR potential of a system is expressed by the symbol E_h (Electrode potential with reference to hydrogen electrode).It is expressed in millivolts. $+E_h$ values indicate oxidizing (Fruit juices: +400 mV) and - E_h values indicate reduced condition (Cheese –100mV).

O-R Potential and Oxygen requirement: As aerobic microbes require oxygen and a high OR potential, the oxidizing condition favours aerobes (bacteria, molds and yeasts). Aerobic bacteria growing under reduced conditions are referred as microaerophiles. Anaerobes need complete exclusion of oxygen. A low OR potential or reducing condition favours anaerobes and facultative anaerobes. Oxidation-reduction systems exhibit resistance to change of potential when the concentrations of oxidant and reductant are close to equal. Fresh milk as ordinarily produced, exhibits a potential at a platinum electrode of between + 0.20 and + 0.30 volt. That dissolved oxygen is a major factor in the establishment of this potential. Milk drawn from udder anaerobically is in reduced state and when such milk is exposed to oxygen, its potential rises to a point such that it is positive. Removal of oxygen by washing with oxygen free gas, remove free oxygen from milk, causes the potential to change in negative direction. Bubbling oxygen through these milks will restore the positive potential.

O-R Potential of plant and animal foods: Plant foods tend to have positive values (+300to 400 mV). Solid meats have low value (-200 mV), whereas minced meats have positive value (+200 mV). Living plants and animals generally have a low OR potential, as they resist oxygen diffusion from outside. The presence of reducing sugars and ascorbic acid in plants and the presence of sulfhydryl group(thiol group) in animals result in low OR potential. Many enzymes, co-enzymes and vitamins also contain reducing groups. A fruit or meat would have aerobic condition only on the surface. Processing of fruits removes reducing substances during filtration and extraction and thereby it can be easily spoiled by yeasts compared to the original juice with pulp. The processing can alter the OR potential of food products and may lead to the spoilage by microorganisms.

4. Nutrient Content of foods

Foods provide a rich habitat for microbes, as most foods contain an abundance of nutrients such as carbohydrates, proteins and lipids and growth factors, minerals and vitamins. Substrate composition of food has a major influence on the micro flora in different foods. Milk has been recognized as an almost indispensable food for mankind. Milk contains all the food constituents, carbohydrate (lactose), protein (casein), fat (butter) required in the human diet and in essentially the proper proportions. Milk supplies exceedingly high quality protein. Milk also supplies liberal amounts of calcium and an excellent source of vitamins, necessary to health.

Foods for energy, growth and multiplication: Sugars are the most commonly used energy source of microbes. Other compounds are alcohols, amino acids, organic acids, starch and cellulose. Most microbes especially fungi require sugar as glucose. Lipolytic and proteolytic microbes use fat and protein respectively when easily utilizable sugars are lacking. They also require sufficient nitrogen and accessory growth substances like vitamins and nucleotides. Most of the food microflora are fastidious (critical) microbes, lack biosynthetic capacity for vitamin production and use B vitamins in meat, ascorbic acid (vitamin C) in fruits.

Molds can utilize many kinds of foods ranging from simple to complex as they possess a variety of enzymes. *Pencillium* species can utilize many kinds of food nutrients. *Penicillium oxalcium, P. islandicum, P. funiculosum* are associated with cereals. *P. crustosum, P. commune, P. echinulatum* are found in lipid and protein rich foods - Cheese, meat and nuts. *P. expansum, P. hirsutum, P. italicum* are found on fruits and vegetables.

5. Inhibitory substances in foods

The presence antimicrobial substances in food prevent the microbial growth and multiplication. They may be -

i) Originally present in food, or

ii) Developed by the growth of microbes or

iii) Added purposely or accidentally.

i) **Originally present in food:** Organic acid naturally present in foods control microbial spoilage. Benzoic acid is naturally present in cranberries, raspberries, plums, cinnamon and cloves and inhibitory to molds and yeasts. Citric acid in cirus fruits, maleic acid in apple and tamarind distrupt substrate transport by alteration of microbial cell membrane permeability. Lysozyme in egg white, lactenin and anticoliform factor in freshly drawn milk are inhibitory to bacteria. The natural antibacterial property of milk was observed and attributed to the activity of enzyme, lysozyme. The lysozyme content of cow's milk is reported as 13 µg per 100 ml. Lactoperoxidase, enzyme present in milk, catalyzes the transfer of oxygen from hydrogen peroxide to other substances and act as antimicrobial substance in milk.

ii) **Developed by the growth of microbes:** The microbes growing in fermented foods may produce substances like acid, alcohol, antibiotics, bacteriocins inhibitory to pathogenic and spoilage organisms. Lactic acid bacteria (LAB) produce lactic acid and also substances like diacetyl and reuterin inhibitory to molds and bacteria. Propionic acid produced by *Propionibacterium shermanii* in Swiss cheese is inhibitory to molds. Alcohol in wine inhibits other competitors of yeast. Bacteriocins are produced by many genera of starter bacteria such as *Lactococcus, Leuconostoc, Pediococcus, Lactobacillus, Bifidobacterium, Propionibacterium.* These cultures are protective to human health and capable of forming substances like bacteriocins, H_2O_2, diacetyl, reuterin by metabolism in dairy foods that inhibit other microorganisms.

iii) **Added purposely/accidentally:** Food Additives, natural and synthetic compounds, are added in food to supply nutrients, to enhance colour, flavour, or texture, and to prevent or delay spoilage. Lemon oil or orange oil when added to grape fruit juice at 0.3% inhibits mold growth and aflatoxin production. Spices and essential oils widely used to impart flavour to foods inhibit microbial growth (Table 1.5). Cinnamon oil at 0.02% concentration inhibited mold growth. Garlic contains about 0.1 percent essential oil, the principal components of which are diallyl disulfide, diallyl trisulfide, and allyl propyl disulfide. Curative powers have been attributed to onions throughout the centuries; they have been recommended for such varied ailments as colds, earaches, laryngitis, animal bites, powder burns, and warts. The onion's characteristic pungency results from the sulfur-rich volatile oil it contains.

TABLE 1.5. PLANT PRODUCTS WITH ANTIMICROBIAL ACTIVITY

PLANT	CLASS	COMPOUND	ACTIVITY
Basil-ocimum	Trepenoid	Essential oil	Bacteria, Salmonella
Turmeric	Trepenoid	Curcumin	Bacteria, Protozoa
Green tea	Flavanoid	Catechin	Shigella, S.mutans
Hops	Phenolic acid	Lupulone	General
Olive oil	Aldehyde	Hexanal	General
Onion	Sulfoxide	Allicin	Bacteria, Candida
Black pepper	Alkaloid	Piperine	Mold, Micrococcus, E.coli
Peppermint	Trepenoid	Methanol	General
Rosemary	Trepenoid	Essential oil	General
Thyme	Thymol	Polyphenol	Bacteria, Mold, Viruses

Since ancient times table salt has been a preservative for fish, ham, and bacon, and sugar has been used to preserve jelly, fruit jams, and fruit preserves. At present more than 2500 food additives of all kinds are known. Other preservative compounds used today include benzoic acid (C_6H_5COOH), sodium benzoate (C_6H_5COONa) and calcium propionate $(C_2H_5COO)_2Ca$. Sulphur dioxide is extensively used to control the growth of undesirable microbes in fruits, fruit drinks, wine, sausages and pickles.

6. Biological structure

Biological structure of foods is the protective covering in plant and animal foods. The inner parts of whole, healthy tissues of living plants and animals are either sterile or low in microbial content. Plants have evolved mechanisms of defense against the invasion and proliferation of microbes. There is a protective covering about the food. Structures such as testa of seeds, shell of nuts, rind or skin- the outer covering of fruits and vegetables, shell on eggs, hide of animals, skin over poultry, scales on fish prevent microbial entry and spoilage. We can also protect them by artificial covering such as wax coating or polyethylene over fruits. The physical protection of food may not only help in its preservation but may also determine the kind, rate and course of spoilage.

These six intrinsic (internal) factors represent nature's way of prevention of microbial spoilage of foods. The contact between foods and microbes is minimized by these factors.

The extrinsic factors affecting microbial growth are temperature of storage, humidity of environment and presence and concentration of gases in the environment. The storage temperature of finished products influences the specific types of microorganisms like psychrotrophs, mesophiles and thermophiles. The selection of storage condition is highly critical in retarding microbial growth. Storage under modified atmosphere is found to have significant advantages in extending shelf life of foods.

QUESTIONS

1. Mention the four aspects of microbial activity that are relevance to processing and preservation of food.
2. What are the various microorganisms that are important in food microbiology?
3. Write a detailed account on disease causing organisms in food.
4. Discuss the various primary sources of microorganisms in foods.
5. Differentiate between homolactic and heterolactic acid bacteria.
6. Explain how food serves as a substrate for microorganisms.
7. Discuss various chief compositional/intrinsic factors affecting the growth of Microbes in food.
8. What is water activity? How does this influence the suceptability of food to spoilage microbes?
9. Write short notes on i) Fermented foods. ii) Lactic acid bacteria. iii) Coliforms. iv) Dust bacteria. v) halophiles. vi) Thermophiles. vii) Psychrophiles. viii) Non acid foods. ix) Water activity. x) Osmophilic yeast. xi) O-R Potential. xii) Inhibitory substances in food.
10. Match the following bacteria with their most characteristic feature

1. *Pediococcus*	-	a. intestinal bacteria
2. *Acetobacter*	-	b. anaerobic spore former
3. *Serratia*	-	c. ferment sugar to lactic acid
4. *Clostridium*	-	d. production of vinegar
5. *E. coli*	-	e. pigmented bacteria

11. Match the following microbes with their source of contamination

1. *Micrococcus*	-	a. soil
2. *Staphylococcus*	-	b. shell fish
3. Hepatitis A	-	c. sea water
4. *Vibrio*	-	d. dust
5. *Clostridium botulinum*	-	e. food handlers

12. Match the following foods with their most characteristic feature

1. Milk	-	a. low water activity
2. Cheese	-	b. ethanol
3. cereals	-	c. high acid food
4. Cranberries	-	d. non-acid food
5. Wine	-	e. benzoic acid

13. Match the following

i) Thermophile	-	a. *Micrococcus*
ii) Thermoduric	-	b. *Bacillus stearothermophilus*
iii) Psychrophile	-	c. *Vibrio parahaemolyticus*
iv) Halophile	-	d. *Pseudomonas*

14. Match the following

i) Soil	-	a. *Vibrio parahaemolyticus*
ii) Sea water	-	b. Hepatitis A virus
iii) Dust	-	c. *Clostridium botulinum*
iv) Shell fish	-	d. *Micrococcus*

2

FOOD PRESERVATION

Agricultural food production is increased by adopting scientific farming methods and another way to increase the availability of food is preventing the food spoilage by adopting food preservation methods. The global agricultural and fish production accounts for 50% of perishable food, which requires preservation. Only half of the perishable foods are properly protected. Application of innovative technologies to promote the production of food is indispensable to meet the demand of emerging populations. It is equally important to protect the stored food from the invasion of pests and diseases. A food saved is a food produced.

Food Processing and Preservation, branch of manufacturing that transforms raw animal, vegetable, or marine materials into tasty, nutritious, and safe food products. The industry has its roots in ancient times, as humans have always needed to obtain food and store a portion for later use. Prehistoric humans may have dried fruits in the sun and stored meat in cold areas, such as ice caves. The modern food processing and preservation industry was born in 1809, when French chef and inventor Nicolas Appert, searching for a better way to provide food for Napoleon's army, devised a method for sterilizing food in tightly sealed glass bottles.

Today a wide variety of methods are available to maintain and enhance the appearance and taste of food. Food processing and preservation methods also create products that are convenient for consumers, such as products that are ready to eat or require minimal preparation and cooking. Combining these methods with modern distribution networks makes seasonal crops available year-round in grocery stores all over the world.

Advantages of food processing

1. Processing of food prevents spoilage of food - e.g. Pasteurized milk.
2. Processing ensures availability of food throughout the year -mango squash.
3. Processing avoids distress sale of produce during over production- e.g. processing tomato to tomato jam.
4. Processing adds value to produce and ensures price stability-cashew nuts.
5. Processed food is equal to fresh and healthy food - e.g. canned pine apple slice.

Food preservation refers specifically to the processing techniques that are used to keep food from spoiling and to permit its future consumption. Spoilage is any change that makes food unfit for consumption, and includes chemical and physical changes, such as bruising and browning; infestation by insects or other pests; or growth of microorganisms, such as bacteria, yeast, and molds. The preserved food should retain palatable appearance, flavour, and texture, as well as its original nutritional value.

Some food preservation techniques destroy enzymes, proteins that are present in all raw foods, which are responsible for the chemical and physical changes that naturally occur after harvesting. Food preservation techniques also help eliminate the moisture or temperature conditions that are favorable for the growth of microorganisms. As they multiply and grow, microorganisms are capable of causing food-borne illness. They also break down foods, producing unpleasant changes in taste, texture, and appearance-changes that we recognize as spoilage. Although people have known about spoilage and some preservation methods to prevent it for centuries, it was only in 1857 that French chemist Louis Pasteur demonstrated the role of microorganisms in the process.

Many kinds of agents are potentially destructive to the agreeable or healthful characteristics of fresh foods. Microorganisms, such as bacteria and fungi, rapidly spoil food. Enzymes, which are present in all raw food, are catalytic substances that promote degradation and chemical changes affecting especially texture and flavour. Atmospheric oxygen may react with food constituents, causing rancidity or colour changes (browning in apples). Equally as harmful are infestations by insects and rodents, which account for tremendous losses in food stocks.

Preservation is a process of inhibiting, retarding and arresting the growth of microorganisms. There is no single method of food preservation that provides protection against all hazards for an unlimited period of time. Canned food stored in Antarctica near the South Pole, for example, remained edible after 50 years of storage, but such long-term preservation cannot be duplicated in the hot climate of the tropics. Besides canning and freezing, other traditional methods of preservation include drying, salting, and smoking. Freeze-drying is a more recent method. Among recent experimental techniques are the use of antibiotics and exposure of foods to nuclear radiation. Food preservation in food industries always involves the use of combination of methods for achieving maximum effectiveness.

The principles involved in food preservation are high temperature, low temperature, dehydration, osmotic pressure, radiation, High Pressure processing (Physical forces) and various chemicals -food preservatives, antibiotics, bacteriocins and food ingredients.

2.1. HIGH TEMPERATURE

The killing of microorganisms by heat is due to thermal denaturation of proteins and enzymes, essential for their metabolic activity and growth. The type of heat treatment depends on kind of organism to be killed, its state and environment during heating, other food preservative methods to be employed and the effect of heat on food. High temperature is widely used in food preservation methods like pasteurization, blanching and canning.

i) Terms used in high temperature food preservation

Thermal death point (TDP) is defined as the temperature necessary to kill all the organisms in a fixed time, usually ten minutes. This is no longer used in food industries as both temperature and time are varied in food processing.

Thermal death time (TDT): The heat resistance of microorganisms is expressed in terms their thermal death time. It is defined as the time required at a certain temperature to kill a stated number of microorganisms under specified conditions. The Thermal death time varies with the type of microorganism and temperature (Table 2.1).

TABLE 2.1 THERMAL DEATH TIME OF FOOD PATHOGENS

ORGANISM	TEMPERATURE (°C)	TIME (minutes)
E. coli	57	30
Staphylococcus aureus	60	19
Salmonella typhi	60	4
Streptococcus thermophilus	75	15
Bacillus anthracis spores	100	2
Bacillus subtilis spores	100	15
Clostridium botulinum spores	100	200

D value (Decimal reduction time) (Fig. 2.1) is defined as the time of heating at a temperature to cause a 90 per cent reduction in the count of viable cells or spores in a substrate (food). D value is constructed from the thermal death time curve and it is 10 minutes for *Bacillus subtilis* at 100°C.

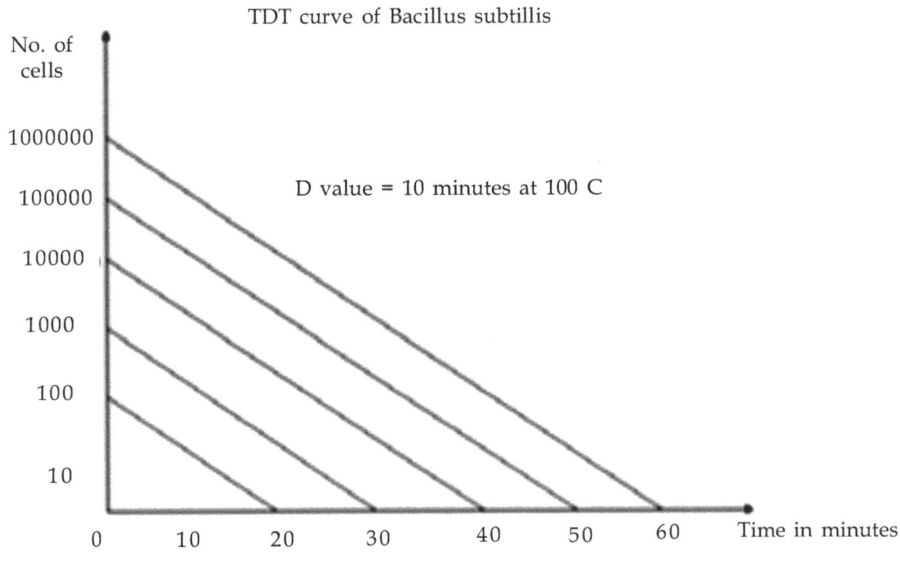

Fig. 2.1 D value

Z Value is defined as increase in temperature (in Fahrenheit) required to reduce the processing time (D value) by ten times. While D value gives the resistance of an organism to a specific temperature, Z value provides information on the relative resistance of an organism to different destructive temperatures. It allows for the calculation of equivalent thermal processes at different temperatures.

For an organism, 100 minutes at 235°F is considered as adequate process with z value 19. What are the temperatures required for 10 minutes and one minute processing time? (Fig. 2.2) (Answer: 254°F (235+19) for 10 minutes and 273°F (254+19)for one minute processing time)

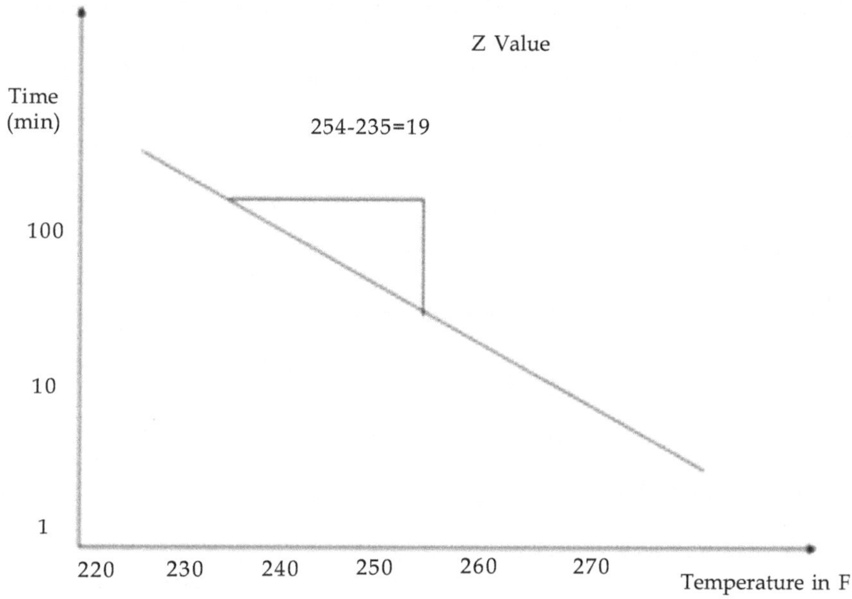

Fig. 2.2 Z value

12 D Concept: Botulism is a very serious type of food intoxication disease caused by *Clostridium botulinum*, an anaerobic, Gram positive, spore forming bacillus. The major source of botulism is improperly processed canned vegetable sausage of low acid content (peas, beans, asparagus sausages). The canning industry has recommended a 12 D treatment for *Clostridium botulinum* spores. It is the heat process required to reduce 10^{12} spores to 1 spore per ml in processed food. As a measure of safety a large initial population of the food poisoning bacterium is assumed to be present.

D value for *Clostridium botulinum* at 250°F (121°C) is 0.21 minutes

12 D value $= 0.21 (\log 10^{-1} - \log 10^{12})$

$= 0.21 \times 12$

= 2.52 minutes.

HIGH TEMPERATURE METHODS

High temperature is widely used in food preservation methods such as pasteurization, blanching and canning.

i) PASTEURIZATION

The process of pasteurization is named after French chemist Louis Pasteur, who devised it in 1885, to inhibit spoilage organisms in wine by applying heat at temperature below its boiling point. The process was later applied to milk and remains the most important operation in the processing of milk. Milk is pasteurized by heating at a temperature of 63°C (145°F) for 30 minutes, rapidly cooling it, and then storing it at a temperature below 10°C (50°F). Beer and wine are pasteurized by being heated at about 60°C (140°F) for about 20 minutes; a newer method involves heating at 70°C (158°F) for about 30 seconds and filling the container under sterile conditions. The device used for pasteurization is called pasteurizer.

Pasteurization is defined as a process of heating a liquid food to a controlled temperature below its boiling point for a specific time, to enhance the keeping quality and destroy harmful microorganisms without changing the composition, flavour, taste or nutritive value of the liquid food.

Objectives of pasteurization: The objectives of pasteurization are two, i) destruction of majority but not all pathogenic and spoilage organisms in liquid foods. ii) Extending the shelf life of food from a microbial and enzymatic point of view. Milk, wine, beer, and fruit juices are all routinely pasteurized.

Pasteurization is most important in all dairy processing. It is the biological safeguard which ensures that all potential human pathogens are destroyed. Extensive studies have determined that heating milk to 63°C (145°F) for 30 minutes or 72°C (161°F) for 15 seconds kills the most resistant harmful bacteria. In actual practice, these temperatures and times are exceeded, thereby not only ensuring safety but also extending shelf life.

Pasteurization types: Low temperature long time (LTLT) process of pasteurization of milk-HOLDING method, was practiced in several countries, required temperatures of about 63°C (145°F) maintained for 30 minutes. Nowadays alternatively, heating to a higher temperature, 72°C (162°F), and holding for 15 seconds - High temperature short time (HTST) process- continuous method, is widely followed. The times and temperatures are those determined to be necessary to destroy the *Mycobacterium tuberculosis* and other more heat-resistant of the non-spore-forming, disease-causing microorganisms such as *Coxiella burnetii* that cause Q fever found in milk. The treatment also destroys most of the microorganisms that cause spoilage and so prolongs the storage time of food.

Ultra-high-temperature (UHT) pasteurization involves heating milk or cream to 138 to 150°C (280 to 302°F) for one or two seconds. Packaged in sterile, hermetically sealed containers, UHT milk may be stored without refrigeration for months. Shelf life is extended to 60–90 days. After opening, spoilage times for ultra-pasteurized products are similar to those of conventionally pasteurized products. The phosphatase enzyme present in fresh milk is destroyed by adequate pasteurization. The absence of the enzyme in phosphatase test indicates proper pasteurization. Pasteurization of some solid foods involves a mild heat treatment, the exact definition of which depends on the food.

Pasteurization methods: a) Holding method (Batch method): Milk is pasteurized in batches in stirred jacketed stainless steel vessels (vat) using steam or hot water. The batch method uses a vat pasteurizer which consists of a jacketed vat surrounded by either circulating water, steam or heating coils of water or steam. In the vat the milk is heated and held throughout the holding period while being agitated. Rapid cooling of pasteurized milk to less than 6°C is necessary to limit the growth of thermophilic bacteria, hence the milk is passed through a cooler. As a modification, the milk may be partially heated in tubular or plate heater before entering the vat. This method is used for milk by-products such as cream and chocolate.

b) Flash method: (Continuous method): Pre-heating, heating, holding and cooling are the four stages in continuous pasteurization (Fig. 2.3). Steel plate heat exchanger is used for heating the liquid foods such as milk and fruit juice. This piece of equipment consists of a stake of corrugated stainless steel plates clamped together in a frame. Gaskets are used to define the boundaries of the channels and to prevent leakage. The heating medium can be vacuum, steam or hot water. The hot pasteurized milk is used for preheating the raw milk and during the step itself get cooled. Continuous process method has several advantages over the vat method, the most important being time and energy saving. The method is conducted in a series of stainless steel plates and tubes, with the hot pasteurized milk on one side of the plate being cooled by the incoming raw milk on the other side. This "regeneration" can be more than 90 percent efficient and greatly reduces the cost of heating and cooling.

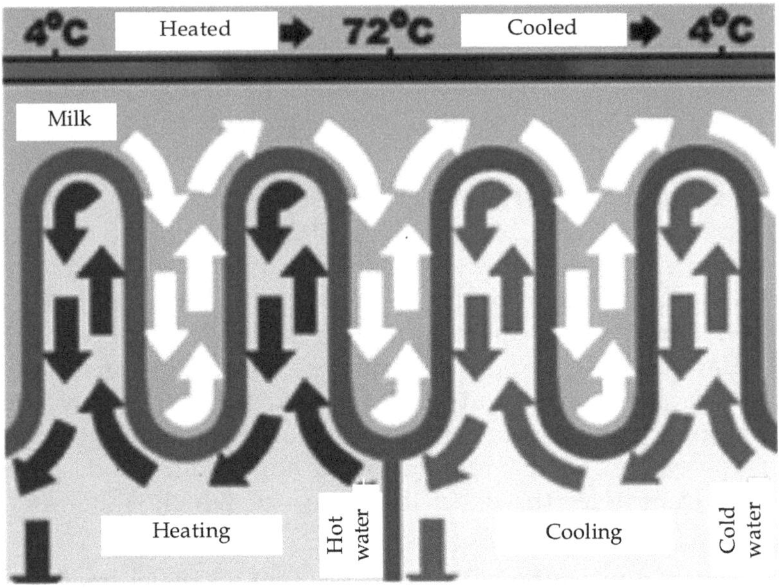

Fig. 2.3 Flash Pasteurization

ii) **BLANCHING**

It is a hot water or steam treatment of immersing raw food (vegetables/fruits) at 63 to 70°C for 2 to 3 minutes before canning.

Objectives

1. Inactivation of enzymes in vegetables that promote degradation and chemical changes affecting especially texture and flavour.
2. Reduction in number (up to 99%) of microorganisms in the food.
3. It helps in fixing the natural / green colour of the vegetables, which makes them attractive in appearance.
4. It softens tissues and facilitates the packing of leafy vegetables in containers.
5. It facilitates displacement of entrapped air in plant tissue.

iii) **CANNING**

Importance: Canning is used to preserve a wide variety of foods, including soups, sauces, fruits, vegetables, juices, meats, fish, and some dairy products. Canning preserves food by heating it in airtight, vacuum-sealed containers. The can is filled with food, and air is pumped out of the space remaining at the top of the can to form a vacuum.

The container is sealed, heated in a cooker called a retort, and then cooled to prevent overcooking of the food inside. This process removes oxygen, destroys enzymes involved in food spoilage, and kills most microorganisms that may be present in the food.

Canned foods are popular because they are already partially prepared and cooked, can be stored without refrigeration for long periods, and are generally low in cost. However, because of the high temperatures required for sterilization, canning affects the colour, texture, flavour, and nutrient content of foods. Fat-soluble vitamins and minerals are barely affected by heat processing, but water-soluble vitamins, especially thiamine, riboflavin, and vitamin C, can leach into canning or cooking water that may later be thrown away during preparation. Up to half of the original content of water- soluble vitamins in a canned product can be lost in this way. Rapid, high-temperature processes generally conserve nutrient content best, as every 4.4°C (18°F) rise in processing temperature yields approximately a ten-fold increase in microbial destruction, with little additional nutrient loss.

Canning is a method of preserving food from spoilage by storing it in containers that are hermetically sealed and then sterilized by heat. Canning is a sterilization process of preserving food by heating and sealing it in airtight containers. Canning eliminates all microorganisms by exposing the food to high temperature (above100°C) for few minutes or seconds to retain their natural flavour. The process of canning is sometimes called sterilization because the heat treatment of the food eliminates all microorganisms that can spoil the food and those that are harmful to humans, including directly pathogenic bacteria and those that produce lethal toxins. Most commercial canning operations are based on the principle that bacteria destruction increases tenfold for each 10°C (18°F) increase in temperature. Food exposed to high temperatures for only minutes or seconds retains more of its natural flavour. In the Flash 18 process, a continuous system, the food is flash-sterilized in a pressurized chamber to prevent the superheated food from boiling while it is placed in containers. Further sterilizing is not required.

History: The process was invented after prolonged research by Nicolas Appert of France in 1809, in response to a call by his government for a means of preserving food for army and navy use. Appert's method consisted of tightly sealing food inside a bottle or jar, heating it to a certain temperature, and maintaining the heat for a certain period, after which the container was kept sealed until use. It was 50 years before Louis Pasteur was able to explain why the food so treated did not spoil: the heat killed the microorganisms in the food, and the sealing kept other microorganisms from entering the jar. In 1810 Peter Durand of England patented the use of tin- coated iron cans instead of bottles, and by 1820 he was supplying canned food to the Royal Navy in large quantities. European canning methods reached the United States soon thereafter, and that country eventually became the world leader in both automated canning processes and total can production. In the late 19th century, Samuel C. Prescott and William Underwood of the United States set canning on a scientific basis by describing specific time-temperature heating requirements for sterilizing canned foods (Fig. 2.4).

Fig. 2.4 Canned Foods

Originally, cans consisted of a sheet of tin-plated iron that was rolled into a cylinder (known as the body), onto which the top and bottom were manually soldered. This form was replaced in the early 20th century by the modern sanitary, or open-top can, whose constituent parts are joined by interlocking folds that are crimped, or pressed together. Polymer sealing compounds are applied to the end, or lid, seams, and the body seams can be sealed on the outside by soldering. The modern tin can is made of 98.5 percent sheet steel with a thin coating of tin (*i.e.,* tinplate). It is manufactured on wholly automatic lines of machinery at rates of hundreds of cans per minute.

Most vegetables, fruits, meat and dairy products, and processed foods are stored in tin cans, but soft drinks and many other beverages are now commonly stored in aluminum cans, which are lighter and do not rust. Aluminum cans are made by impact extrusion; the body of the can is punched out in one piece from a single aluminum sheet by a stamping die. This seamless piece, which has a rounded bottom, is then capped with a second piece as its lid. The tabs used in pop- top cans are also made of aluminum. Bimetal cans are made of aluminum bodies and steel lids.

RAW FOOD

↓

CLEANING IN WATER & GRADING

↓

FOOD CUT / SLICED / PEELED

↓

BLANCHING AT 63-75°C

↓

FILLING CAN WITH SOLID& LIQUID (BRINE/SYRUP)

↓

STEAM BATH & SEALING

↓

STERILIZTION AT 116°

↓

CANS COOLED, LABELLED & PACKAGED

Fig. 2.4. FLOW CHART OF CANNING

STAGES IN CANNING

Canneries are usually located close to the growing areas of the product to be packed, since it is desirable to can foods as quickly as possible after harvesting. The canning process itself consists of several stages:

1. Cleaning and grading the raw food material.
2. Blanching
3. Filling the containers, usually under a vacuum; closing and sealing the containers;

4. Sterilizing the canned products and cooling, labeling and warehousing the finished goods.

1. **Cleaning and grading:** Cleaning usually involves passing the raw food through tanks of water or under high-pressure water sprays, after which vegetable or other products are cut, peeled, cored, sliced, graded, soaked, pureed, and so on.
2. **Blanching:** Almost all vegetables and some fruits require blanching by immersion in hot water or steam; this process softens the vegetable tissues and makes them pliable enough to be packed tightly, while also serving to inactivate enzymes that can cause undesirable changes in the food before canning. Blanching also serves as an additional or final cleansing operation.
3. **Filling and Sealing:** The filling of cans is done automatically by machines; cans are filled with solid contents and, in many cases, with an accompanying liquid (often brine or syrup) in order to replace as much of the air in the can as possible. The filled cans are then passed through a hot-water or steam bath in an exhaust box; this heating expands the food and drives out the remaining air; Immediately after the cans are exhausted, they are closed and sealed; a machine places the cover on the can, and the curl on the can cover and the flange on the can body are rolled into position and then flattened together. The thin layer of sealing compound originally present in the rim of the cover is dispersed between the layers of metal to ensure a hermetic seal.
4. **Sterilization:** The sealed cans are then sterilized; *i.e.*, they are heated at temperatures high enough and for a long enough time to destroy all microorganisms (bacteria, molds, yeasts) that might still present in the food contents. The heating is done in high-pressure steam kettles, or cookers, usually using temperatures around 240°F (116°C). The cans are then cooled in cold water or air, after which they are labelled. After sealing, heat sterilizing and cooling the can, the contraction of the contents produces a partial vacuum within the container. Certain products are vacuum-packed, whereby the cans are mechanically exhausted by specially designed vacuum- can sealing machines. Canning preserves most of the nutrients in foods. Proteins, carbohydrates and fats are unaffected, as are vitamins A, C, D, and B_2. The retention of vitamin B_1 depends on the amount of heat used during canning. Some vitamins and minerals may dissolve into the brine or syrup in a can during processing, but they retain their nutritive value if those liquids are consumed.

The canning process can be illustrated by the example of green beans (*Phaseolus vulgaris* L.). After arrival at the processing plant, the beans are conveyed to size graders. Graders consist of revolving cylinders with slots of various diameters through which the beans fall onto conveyers. The conveyers carry them to snipping machines, where their tips and stems are cut off. The snipped beans then pass over inspection belts, where defective beans are removed. Smaller beans are canned as whole beans, while larger beans are cut crosswise by machine into various lengths. Some smaller beans are cut lengthwise and marketed as French-cut beans. Both the small whole and cut beans are blanched for 1.5 to 2 minutes in 82°C(150°F) water and mechanically packed in cans. The cans are then filled with hot water and dry salt or with brine, steam-exhausted for approximately five minutes, and sealed while hot or with steam flow. Depending on the size of the can, they are heat-processed for various periods of time-from 12 minutes at 120°C (248°F) to 36 minutes at 116°C (240°F). The cans are cooled to room temperature, labelled, and packaged for storage or immediate distribution.

2.2. LOW TEMPERATURE

Low-temperature storage as a preservation method probably began when prehistoric humans stored meat and other foods in ice caves. Low temperature is widely used in food preservation as it retards the chemical reactions and activity of food enzymes; slows down or stops the growth and activity of spoilage and pathogenic microorganisms. However, mechanical refrigeration and large-scale freezing are relatively recent innovations. Mechanical refrigeration was pioneered by American inventor John Gorrie in 1842, but a mechanical refrigeration system suitable for widespread commercial use was not developed until the 1870s. American inventor Clarence Birdseye developed procedures, equipment, and packaging for quick-freezing in the 1920s.

Advantages of low temperature

1. Low temperature preserves food without any adverse effect on nutritional value, flavour, colour and texture of food.
2. It controls the rate of chemical and enzymatic changes in food due to spoilage organisms and endogenous enzymes.
3. It makes the processing of foods easy-peeling of fruits and vegetables for canning.

Effect of low temperature on microbial activity

The rate of growth of microorganisms decrease in low temperature, but slow metabolic activity may continue. The complete death of all microorganisms does not occur in low temperature and when the food is thawed, there can be rapid multiplication of microorganisms. Low temperature is widely used in food preservation methods like cool storage, refrigeration and freezing.

Low Tempertaure Methods

i) **Cool storage**

Bulk storage of fruits and vegetables in cool environment created by air cooler or water spray is widely practiced in major fruits and vegetable growing areas. Apples and grape fruits store well at 10°C and 11°C respectively. Banana and tomato are stored at 13°C to slow down natural ripening. Potato and cabbage are stored at 15°C. Loss of crispness of fruits and vegetables, loss of flavour, loss of nutrients- sugars in sweet corn are few problems associated with cool storage.

ii) **Refrigeration**

Refrigeration is advantageous because it does not cause chemical or physical changes to food. Refrigeration or chilling is used for storage of foods, meats, fish, eggs, milk, some fruits, and some vegetables, at -2°C to 10°C for few days to weeks or months. Apples can be stored without any reduction in quality for six months at 2.8°C and 87 % relative humidity. The air circulated in storage room must not be too dry (less than 50% relative humidity) or too moist (more than 95% relative humidity). The amount and proportion of gases in storage atmosphere influence preservation by refrigeration.

iii) **Freezing**

Storage of foods in frozen condition is an important method of food preservation. Freezing allows foods to be stored for longer periods than refrigeration because it inhibits enzyme activity and microbial growth to a greater degree. Although prehistoric humans stored meat in ice caves, the food-freezing industry is more recent in origin than the canning industry. The freezing process was used commercially for the first time in 1842, but large-scale food preservation by freezing began in the late 19th century with the advent of mechanical refrigeration. Frozen foods have the advantage of resembling the fresh product more closely than the same food preserved by other techniques. Frozen foods also undergo some changes, however. The greatest disadvantage of freezing is that the water in food expands and forms ice crystals. The ice crystal formation disrupts the structure of plant and animal cells, giving frozen food a softer texture after thawing. Newer technologies in which freezing occurs more rapidly help minimize this problem: faster freezing means that smaller ice crystals form, resulting in less damage to cells.

Effect of freezing on microbes: Freezing preserves food by preventing microorganisms from multiplying and retarding the action of food enzymes. Because the process does not kill all types of bacteria, however, those that survive reanimate in thawing food and often grow more rapidly than before freezing. Freezing usually results in a considerable reduction in number of viable microorganisms due to its lethal and sub-lethal effects.

Lethal effect: Rapid cooling of microbial cells in food from an optimal temperature (room temperature) to freezing temperature can cause cold shock and result in death of microbial cells. This is due to alteration of lipids in cell membranes and damage to permeability of cells.

Sub-lethal effect: Freezing causes the cytoplasmic water in microbial cell to expand and tends to disrupt the cell structure by forming ice crystals. This results in ice injury or cryo-injury of microbial cells in food. The freeze injured cells can be recovered if repair time is permitted.

Vegetables are blanched or heated in preparation for freezing to ensure enzyme inactivity and thus to avoid degradation of flavour. Blanching has also been proposed for fish, in order to kill cold-adapted bacteria on their outer surface. In the freezing of meats various methods are used depending on the type of meat and the cut. Pork is frozen soon after butchering, but beef is hung in a cooler for several days to tenderize the meat before freezing.

Methods of freezing

1. **Air freezing:** In this method cold air is blown across the food. In slow air freezing -15 to -23°C cold air is blown for three hours and in quick air freezing-18 to -34°C cold air is blown for 30 minutes. In one type of freezer used for packaged foods, the packages are transported mechanically on a conveyor belt through an air blast, which produces temperatures as low as -40° C. In quick- freezing the ice crystals are smaller, producing less cell damage than in the slowly frozen product. The quality of the product, however, may depend more on the rapidity with which the food is prepared and stored in the freezer than on the rate at which it is frozen. Some solid foods that are frozen slowly, such as fish, may, upon thawing, show a loss of liquid called drip; some liquid foods that are frozen slowly, such as egg yolk, may become coagulated.

2. **Indirect air freezing:** In this method the food does not come in contact with refrigerant but come in contact only with cold surface maintained at -18 to -45°C. In a widely used freezer called the plate freezer, the packages are in contact with hollow metal plates containing a refrigerant and are subjected to pressure in order to increase the rate of freezing. This method of preservation is the one most widely used for a great variety of foods, including bakery goods (both ready to eat, and to be cooked when desired), soups, and precooked complete meals.
3. **Direct immersion freezing:** In this method, food is immersed in refrigerant, fish in brine. Another type of freezer, used in the freezing of concentrated orange juice, contains a secondary refrigerant, such as calcium-chloride brine, as a spray or a bath for the cans at temperatures of -29° C. Food are also immersed in liquid nitrogen or sprayed with liquid nitrogen.

Because of the high cost of refrigeration, frozen food is comparatively expensive to produce and distribute. High quality is a required feature of frozen food to justify the added cost in the market.

2.3. DEHYDRATION

Drying and Dehydration

Drying has been used to preserve food by cultures throughout the world since prehistoric times, when people learned that dried foods-for example, fruits left out in the sun-remain wholesome for long periods. In the case of meat and fish, other preservation methods, such as smoking or salting, which yielded a palatable product, were generally preferred. Commercial dehydration of vegetables was initiated in the United States during the American Civil War but, as a result of the poor quality of the product, the industry declined sharply after the war. In modern times, the dried foods industry greatly expanded after World War II (1939-1945) but remains restricted to certain foods, including milk, soup, eggs, fruits, yeast, some meats, and instant coffee, that are particularly suited to the process.

Purpose of Drying

1. Preservation of foods.
2. Decreasing the weight and bulk of the food thereby reducing transportation and storage costs.

3. Production of convenience foods such as instant coffee and instant mashed potatoes, in which cases brewing or cooking steps are completed before the products are dried.

Although both these terms- drying and dehydration, are applied to the removal of water from food, to the food technologist drying refers to natural desiccation, such as spreading fruit on racks in the sun, and dehydration designates drying by artificial means, such as a blast of hot air.

Three basic methods of drying are used today:

i) **Sun drying:** A traditional method in which foods dry naturally in the sun. Fruits such as raisins, figs, apricots, dates, pears and peaches and salted fish, meat and vegetables are spread out on trays and turned on their sides during drying. By sun drying the moisture content of these foods is lowered to 15% and stable for a short period.

ii) **Mechanical driers:** Hot air drying, in which foods are exposed to a blast of hot air or drying by direct contact with a heated surface or by application of energy from a radiating microwave source. Present-day dehydration techniques include the application of a stream of warm air to vegetables. Liquid food is dehydrated usually by spraying it as fine droplets into a chamber of hot air, or occasionally by pouring it over a drum internally heated by steam.

iii) **Freeze-drying** or lyophilization is used to preserve products containing water or solvents. In freeze drying, moisture in the food is frozen and then sublimed to vapour. The food is placed in a vacuum chamber to draw out the water. There, water escapes from the food by sublimation, a process in which ice changes from a solid directly to a vapour without first becoming a liquid. In freeze-drying a high vacuum is maintained in a special cabinet containing frozen food until most of the moisture has sublimed. Protein foods such as meat are of good quality only if freeze-dried. Freeze-dried foods retain their original colour, taste, original shape, flavour, texture, and nutrients upon rehydration but must be packaged in moisture-proof, hermetically sealed containers. Weight reduction will be up to 92%. Freeze-drying is an expensive process used for such products as instant coffee, tea, dried soup mixes, strawberries, mushrooms, meat and meat products, dairy products cheese, yoghurt and shrimp.

Removing the water preserves food because microorganisms need water to grow and food enzymes cannot work without a watery environment. Removal of water offers excellent protection against the most common causes of food spoilage. Microorganisms cannot grow in a water-free environment, enzyme activity is absent, and most chemical reactions are greatly retarded. This last characteristic makes dehydration preferable to canning if the product is to be stored at a high temperature. In order to achieve such protection, practically all the water must be removed. The food then must be packaged in a moisture-proof container to prevent it from absorbing water from the air. For this reason a hermetically sealed can is frequently used to store dry foods. Such a can offers the further advantage of being impervious to external destructive agents such as oxygen, light, insects, and rodents.

Vegetables, fruits, meat, fish, and some other foods, the moisture content of which averages as high as 80 percent, may be dried to one-fifth of the original weight and about one-half of the original volume (Table 2.2). Removing the water also decreases the weight and volume of foods, thereby reducing transportation and storage costs. The disadvantages of this method of preservation include the time and labour involved in re-hydrating the food before eating. Further, reconstituting the dried product may be difficult, because it absorbs only about two-thirds of its original water content; this phenomenon tends to make the texture tough and chewy.

TABLE 2.2 REDUCTION IN MOISTURE CONTENT OF FOODS

FOOD	MOISTURE%	
	BEFORE DRYING	AFTER DRYING
MILK	87	50
EGG	74	30
CHICKEN	60	20
BOILED POTATO	80	40
APPLE JUICE	86	60

2.4. OSMOTIC PRESSURE

Plasmolysis: In solutions having high osmotic pressure protoplasmic water will leave the microbial cells and cells get plasmolysed. In the plasmolysed cells there is a gap between cell membrane and cell wall. These cells shrink and get killed due to loss of water. This plasmolysis principle is practically applied in the preservation of jam, jellies, condensed milk by adding sugar and in meat, beef, sea foods and pickles by the use of salt. Food preservation is achieved by the addition of solutes like sugar or salt. Since ancient times table salt has been a preservative for fish, ham, and bacon, and sugar has been used to preserve jelly, fruit jams, and fruit preserves.

Salting and Sweetening: Salting of fish and pork has long been practiced, using either dry salt or brine (salt solution). Salt enters the tissue and in effect binds the water, thus inhibiting the bacteria that cause spoilage. Sugar, a major ingredient of jams and jellies, is another preservative agent. For effective preservation, the total sugar content should make up at least 65 percent of the weight of the final product. Sugar, acts in much the same way as salt, inhibits bacterial growth after the product has been heated.

2.5. IRRADIATION

There are many methods of food preservation, of which irradiation is a method of using electromagnetic waves. Irradiation of food, an eco-friendly approach to preserve food from spoilage, has now emerged as a novel technique to store food for longer period.

Radiation may be defined as the emission and propagation of energy through space or a medium. The electromagnetic radiations such as radio waves, microwaves, infrared rays, UV rays, X rays, gamma rays are widely used in different fields. Among the electromagnetic radiations, radio waves are the least energetic and gamma rays are the most penetrating.

Ionizing and non-ionizing radiations: Radiations of primary interest in food preservation are ionizing radiations. Ionizing radiation is the emission and propagation of energy through a medium with radiation wave length of 2000 Å(200 nm) or less like x rays, gamma rays, alpha particles, beta particles and cosmic rays. Ionizing radiations have sufficient energy to ionize molecules in their path. Among the ionizing radiations, gamma rays are widely used in food preservation / food irradiation, because they can penetrate deeply up to half metre into food packed in metallic containers. X rays are not used in food preservation as they are less energetic than gamma rays and costly to produce. UV rays are non-ionizing radiations (wave length 2600 Å) absorbed by proteins and nucleic acids. It has poor penetrating capacity and used for food surface applications like to treat surfaces of baked fruit cakes before wrapping. UV light may catalyze oxidative changes in food, that lead to rancidity, discolourations and other reactions.

Food irradiation is a process in which food products are exposed to a controlled amount of radiant energy to kill harmful bacteria, insects and parasites and prevent food spoilage. Because of growing concern over the use of potentially toxic chemicals as fumigants, the Food and Drug Administration (FDA) approved the use of ionizing radiation as a substitute. Irradiation delays ripening of fruits and vegetables; inhibits sprouting in bulbs and tubers; disinfests grain, cereal products, fresh and dried fruits and vegetables of insects; and destroys bacteria in fresh meats.

The FDA first approved irradiation for use on wheat and wheat flour in 1963, and later approved its use on white potatoes, spices, pork, some fresh produce (onions, tomatoes, mushrooms, and straw berries), and poultry. The irradiation of fresh fruits and vegetables, herbs and spices, and pork was approved in 1986. In 1997, in response to several food-borne illness outbreaks and increasing public concern over the safety of the food supply, irradiation was approved for use on poultry products. In 1999, irradiation was approved to curb pathogens in raw meats including ground beef, steaks, and pork chops. Irradiation is also used to preserve some meals eaten by astronauts during long-term space missions. Irradiated foods were used by U.S. astronauts and by Soviet cosmonauts. The irradiated food is safe for consumption. Public concern over the safety of irradiation has limited its full-scale use.

Chemical method of food preservation is widely followed in food industry which has many drawbacks. The treated chemical may leave residue in the food and may contaminate the environment. Often they cause allergy to some consumers. Moreover food spoilage microbes have the capacity to develop resistance to these chemical molecules. Hence higher doses have to be applied to eliminate the offending organisms in the subsequent generations. Because of growing concern over the use of potentially toxic chemicals as fumigants, the Food and Drug Administration approved the use of ionizing radiation as a substitute. The use of ionizing radiation for microbial decontamination of foods offers a promising alternative to other physical and chemical methods of preservation. Irradiation kills most bacteria, molds, and insects that may contaminate food. Irradiation also delays the ripening of fruits and sprouting of vegetables, permitting produce to be stored for longer periods of time. Because irradiation involves minimal heating, it has very little effect on the taste, texture, and nutritive value of food.

Ultraviolet radiation treatment: It is one the non-thermal preservation method used for fruit and vegetable juices. Typically, the wavelength of UV processing ranges from 100-400 nm. This range may be further subdivided in to UV-A (315-400 nm), UV-B (280-315 nm), UV-C (100-280 nm). UV-C is considered germicidal range since it effectively inactivates bacteria and viruses. UV light kill or interrupt bacterial replication due to the formation of thymine dimers in bacterial chromosome. Short wave UV-C radiation is lethal to most microorganisms and can be applied to render safe food products. UV radiation processing can potentially provide more ideal fresh foods.

Radiation source and units: Food irradiation involves the use of either high-speed electron beams or high-energy radiation with wavelengths smaller than 200 nanometres, or 2000 angstroms (e.g., X rays and gamma rays). These rays contain sufficient energy to break chemical bonds and ionize molecules that lie in their path. The two most common sources of high-energy radiation used in the food industry are cobalt-60 and cesium-137. For the same level of energy, gamma rays have a greater penetrating power into foods than high-speed electrons. The energy possessed by an electron is called an electron volt (eV). One eV is the amount of kinetic energy gained by an electron as it accelerates through an electric potential difference of one volt. It is usually more convenient to use a larger unit such as megaelectron volt (MeV), which is equal to one million electron volts. The unit of absorbed dose of radiation by a material is denoted as the gray (Gy), one gray being equal to the absorption of one joule of energy by one kilogram of food. A Kilogray (KGy), of energy absorbed by a litre of water in the form of heat will raise the temperature by 2.4°C. The dose of irradiation to be given depends on the food item and the purpose for which it is given (Table 2.3).

TABLE 2.3. IRRADIATION DOSAGES FOR FOODS

DOSE (KGy)	FOOD ITEM	PURPOSE
0.05-0.15	Potato onion, garlic	To prevent sprouting of tubers
0.25-0.75	Fruits	To delay ripening of fruits
0.20-1.0	Cereals and dry fruits	Prevents insect infestation and development of grubs into adults
0.30-1.0	Beef, mutton, pork	Inactivate worms and parasitic nematodes
1.0-2.5	Fish and meat in ice	Prolongs life by 2-3 times
3.0-10.0	Various foods	Lethal to all disease causing Gram negative bacteria
30.0	Frozen foods	Maximum dose

Food irradiation technology is already half a century old and more researched than any other method of food preservation. To begin with, there were so many doubts about the wholesomeness of irradiated food. It was feared that the food undergoes chemical changes that may render it unfit for human consumption and wrongly believed to become carcinogenic. Loss of vitamins and genetic changes in irradiated food were also suspected. Much of the data reviewed by FAO and WHO, found no adverse effect with irradiated food. However they have to be stored and distributed chilled (at 5°C) to prevent subsequent spoilage. Thus in most of the cases irradiation is better than chemical treatment.

Irradiation of meat is accomplished by exposing meat to high-energy ionizing radiation produced either by electron accelerators or by exposure to gamma-radiation. The pasteurization of fresh fish using low-level dosages of ionizing radiation may extend the shelf life of the product up to several weeks. The sensory and nutritional characteristics of the fish are unaffected at these low levels of radiation. Irradiated products are virtually identical in character to non-irradiated products, but they have significantly lower microbial contamination. Irradiated fresh food products still require refrigeration and packaging to prevent spoilage, but the refrigerated storage life of these products is greatly extended.

A FAO-IAEA-WHO experts committee on wholesomeness of irradiated foods in 1980 has accorded an unconditional clearance to the process for all types of food treated with an overall average dose of 10 KGy or less as this method introduces no special nutritional or toxicological problems. Irradiated food emits no abnormal amount of irradiation after treatment. Export of fresh mango fruits from India to Western countries was not permitted owing to the attack of pests, but irradiation as a quarantine measure opened up the ways for export of Indian mangoes to these countries

Methods of Irradiation

Radurization or radiation pasteurization refers to the application of small amounts of beta or gamma rays to foods to increase their storage time. Radurization is equivalent to pasteurization to enhance the keeping quality and shelf life of fresh meat, poultry, sea food, fruits, vegetables and cereals by causing reduction in spoilage microbes by radiation. **Radurization** is as effective as heat pasteurization in killing food-spoilage microorganisms. Irradiation of meat is accomplished by exposing meat to high-energy ionizing radiation produced either by electron accelerators or by exposure to gamma- radiation-emitting substances such as cobalt-60 or cesium-137. Irradiated products are virtually identical in character to non-irradiated products, but they have significantly lower microbial contamination. **Radicidation** is equivalent to radiation bactericidation that reduces the non-spore forming pathogens other than viruses and present no health hazard. It is effective on pre-packaged food products (raw poultry, spices and condiments) and eliminates the possibilities of cross contamination. **Radappertzation** is radiation sterilization or commercial sterility of canned foods to eliminate endospores and exotoxin of *Clostridium botulinum*. (Table 2.4).

TABLE 2.4 METHODS OF IRRADIATION

METHOD OF RADIATION	EFFECT	DOSE (KGy)
1. Radurization	Equals to pasteurization	0.75-2.5 KGy
2. Radicidation	Bactericidation- destruction of nonspore forming non-viral microorganisms	2.5-10.0 KGy
3. Radappertzation	Complete Sterilization	30.0-40.0 KGy

Food irradiation procedure

Fig. 2.5 Radura Symbol

Food is packed in containers and moved by a conveyer belt into a shielded room. There the food in containers is briefly exposed to a radiate energy source - gamma rays or X rays. These high- energy rays are strong enough to break chemical bonds, destroy cell walls and cell membranes, and break down deoxyribonucleic acid (DNA), the substance that carries genetic information in bacteria, other pathogens and insects. These organisms die or unable to reproduce. The food is virtually left unchanged, but the number of harmful bacteria, fungi, and parasites is reduced and may be eliminated. The irradiated foods are labelled as 'treated with radiation' or 'treated by radiation' and the international symbol for irradiation RADURA (Fig. 2.5) is pasted on the packing. Very special care has to be taken while handling, storing, using and disposing of radioactive material. During these operations, extreme care has to be taken that it is not mis-utilized, spilled or polluted.

Irradiation effect on microbes: Irradiation of foods eliminates organisms like *Salmonella, Listeria, Campylobacter, Shigella, Vibrio* and *Aeromonas* which are very sensitive to irradiation. Gram negative bacteria are more sensitive to irradiation than Gram positive bacteria. The bacteria most sensitive to radiation belong to *Pseudomonas* and *Flavobacterium* group. Spore formers are more resistant than non-spore formers. *Bacillus sp, Clostridium botulinum* endospores are more resistant. Lactobacilli, Micrococci, *Streptococcus faecalis, Streptococcus faecium* are resistant vegetative forms. *Micrococcus radiodurans* is the most radio resistant bacterium.

Fish having an ice storage life of 7-10 days and lamb and buffalo meat in nonfrozen state (3°C) having 7-14 days storage life can be increased by 2-3 times by irradiation. This is achieved by suppressing the growth of organisms like *Pseudomonas, Proteus, Aeromonas* and *Achromobacter* in animal products.

Biological effects of irradiation: Irradiation has both direct and indirect effects on biological materials. The direct effects are due to the collision of radiation with atoms, resulting in an ejection of electrons from the atoms. The indirect effects are due to the formation of free radicals (unstable molecules carrying an extra electron) during the radiolysis (radiation-induced splitting) of water molecules. The radiolysis of water molecules produces hydroxyl radicals, highly reactive species that interact with the organic molecules present in foods. The products of these interactions cause many of the characteristics associated with the spoilage of food, such as off-flavours and off-odours.

Advantages and Disadvantages

1. Irradiation is a safe, cheap, convenient, eco-friendly method of preservation, but suitable only for bulk handling of huge produce.
2. The food can be first frozen, packed and then irradiated little before shipping.
3. Irradiation has to be done only once, whereas chemical treatment has to be repeated.
4. Even though capital investment is huge (around Rs. 3 crores), compared to freezing the energy requirement for irradiation is low. Irradiation needs only one fifth of the cost for freezing.
5. Milk, butter and fatty meat are not amenable to irradiation

2.6. HIGH PRESSURE PROCESSING OF FOODS

High pressure processing (HPP) or High hydrostatic pressure (HHP) or ultra high pressure processing (UHP) is a non-thermal technique that subject liquid and solid foods with or without packaging to pressures between 100 and 800 MPa. It is gaining popularity with food processors due to increased demand by consumers for improved nutritional and sensory characteristics of food without loss of fresh taste. Application of HPP as a method for microbial inactivation has stimulated considerable interest in food industry.

Vegetative forms of bacteria, yeasts and mold are most pressure sensitive and are inactivated at 400-600 MPa. Effects of HPP on microbial cells include an increase in permeability of cell membranes, inhibition of enzymes vital for survival and reproduction of spoilage and pathogenic bacterial cells. The effect is less pronounced on Gram-positive bacteria than on Gram-negative bacteria. The response of microbial cell to pressure treatment namely, inactivation and survival response depends on pressure level (200 to 700MPa), holding time (millisecond to 1200 seconds-20 minutes), temperature (0 to 100°C) and composition. Milk, jam, fruit juices, vegetable juices, potato cubes, chopped tomato, minced meat and olive oil are processed using HPP.

2.7. CHEMICAL PRESERVATIVES

Preservatives are divided into two classes

Class I: Includes those which can be added to any food, generally on a "no limit" basis. e.g. Acetic acid, Vinegar, Dextrose, Edible oils, Honey, Spices, Common salt and Sugar. Many food additives are approved by CAC as GRAS - Generally Recognized As Safe, and they are exempt from approval because their extensive use in the past has not shown any harmful effects in human beings.

Class II: Includes the following chemicals or their sodium, potassium, calcium salts, benzoic acid, methyl propyl parahydroxy benzoic acid, lactic acid, propionic acid, nitrates, nitrites, sodium diacetate, sorbic acid, sulphur dioxide, sulphides and acid Ca_3PO_4.

ORGANIC ACIDS: Vinegar (acetic acid), because of its high acidity, acts as a preservative in pickling relishes and other foods that have been heated. Fermentation caused by certain bacteria, which produce lactic acid, is the basis of preservation in sauerkraut and fermented sausage. The preservative compounds used today include sodium benzoate (C_6H_5COONa), calcium propionate ($C_2H_5COO_2$ Ca) and benzoic acid (C_6H_5 COOH). Sodium benzoate, restricted to concentrations of not more than 0.1 percent, is used in fruit products to protect against yeasts and molds. Calcium propionate may be added to baked goods to inhibit mold.

INORGANICS: Sulfur dioxide is permitted in most states, the chemical preservative helps to retain the colour of dehydrated foods. Chemicals used in germicidal ice to preserve fish are hypochlorites, chloromine, hydrogen peroxide, ozone, sodium nitrate, and antibiotics. Immersion in water glass, a solution of sodium silicate is a successful method of preservation of eggs. The water glass solution is inhibitory because of its alkalinity. Other inhibitory chemicals used are borates, permanganates, benzoates, salicylates and formates.

Antibiotics such as chlortetracycline, oxytetracycline and chloromphenicol are used in meat to prolong the storage life at chilling or higher temperatures.

Nisin is the best known bacteriocin most frequently studied anti-microbial, low molecular weight protein produced by *Lactococcus lactis* subsp. *lactics*. It is widely used in various foods such as dairy products, canned foods, meat products, etc. Nisin inhibits a wide range of Gram positive bacteria, particularly those that produce spores. Thus within the Gram positive bacteria, nisin inhibits certain strains of species of *Staphylococcus, Streptococcus, Enterococcus, Micrococcus, Lactobacillus* and the majority of spore forming species of *Clostridium* and *Bacillus* with the spores being more sensitive than the vegetative cells. The strains of the food pathogen, *Listeria monocytogenes* are also sensitive to nisin. It was reported that nisin used in combination with a chelating agent EDTA exhibited a bactericidal effect towards Gram negative bacteria, *Salmonella* and *E.coli*. In addition, nisin has been used to inactivate hemophilic spoilage organisms in canned foods and to extend the shelf life of milk and dairy products. In recent years, *in vitro* antagonistic effect of nisin was tested on enterococci, staphylococci, isolated from faecal samples.

Mode of Action

The point of action of nisin is the cytoplasmic membrane and it causes disruption either resulting in leakage of essential cellular material such as ATP from the cell or loss of the proton motive force or in more severe cases lyses. In cell free systems, nisin - inhibited peptidoglycan synthesis catalyzed by enzymes from *Bacillus stearothermophilus* or *E.coli*. The use of nisin to prevent gram positive bacteria contaminating fermentations carried out by nisin- insensitive micro-organisms such as Gram negative bacteria, yeasts or fungi could be applicable in other industrial fermentations such as the production of single cell protein, organic acids, polysaccharides, amino acids and vitamins. Its use as a processing aid to control contaminating bacteria in industrial processes other than fermentations is also being considered. Toxicity studies carried out with levels of nisin far in excess of those that would be used in food indicate that nisin is non-toxic (Table 2.5). Nisin has been shown to be rapidly inactivated in the intestine by digestive enzymes, and nisin cannot be detected in the saliva of humans ten minutes after consuming liquid containing nisin.

TABLE 2.5 BACTERIOCINS IN FOOD PRESERVATION

BACTERIOCIN	ORGANISM PRODUCING	FOODS PRESERVED
NISIN	*Lactococcus lactis* subsp. *lactics*	Milk products
LACTIN	*Lactococcus lactis* subsp. *lactics*	Milk products
LACTOCIN	*Lactococcus lactis* subsp. *sake*	Fermented sausages
CARRIOCIN	*Carrobacterium* sp	Fish
VARIACIN	*Micrococcus variaus*	Fermented meat

Food additives are chemicals that are added to food in small amounts. Direct additives are added deliberately during processing to make food look and taste better, maintain or improve nutritive value, maintain freshness, and help in processing or preparation. Some additives help preserve food by preventing or slowing chemical changes and the growth of microorganisms in food. As many as 3000 substances are approved by the Food and Drug Administration (FDA) for use as direct additives. An additional 10,000 substances are present in foods as indirect additives. These substances enter food incidentally during handling or from processing equipment or packaging.

Food additives have been used for thousands of years. The salts and other chemicals used in curing are additives, and before the advent of canning and mechanical refrigeration, chemical additives were the only means of preservation available. Additives were not limited to use as preservatives, however. People in ancient Rome added certain chemicals to wine and cooked vegetables to improve the colour of these foods. Other examples of additives that have been used since ancient times include yeast and baking powder used as leavening in baked goods.

Another widely used preservation method is **smoking**, which frequently is applied to preserve fish, ham, and sausage. The smoke is obtained by burning hickory or a similar wood under low draft. In this case, some preservative action is provided by such bactericidal chemicals in the smoke as formaldehyde and creosote, and by the dehydration that occurs in the smokehouse. Weight is lost as moisture evaporates during smoking. Smoking usually is intended to flavour the product as well as to preserve it. Smoked fish is considered as delicious. Soaking fish in strong salt solution (brine), before smoking gives a good surface texture and retard surface spoilage.

2.8. FOOD STORAGE AND PACKAGING

Controlled Atmosphere Storage

Fruits and vegetables are sometimes stored in sealed warehouses where temperature and humidity are closely controlled, and perhaps most importantly, the composition of gases in the atmosphere is altered to minimize spoilage. The composition of atmosphere is controlled by the introduction of carbon dioxide or ozone. Eggs and chilled beef are stored well at 2.5% and 10% carbon dioxide concentrations. Usually, the concentration of oxygen is reduced, the concentration of carbon dioxide is increased, and ethylene, a gas naturally produced by plants that accelerates ripening, is removed from the atmosphere.

A method being studied by technologists involves the preservation of fruits and vegetables through the immediate anaerobic treatment of such foodstuffs with gases such as carbon dioxide, carbon monoxide, and nitrogen. This controlled environment helps slow the enzymatic reactions that eventually lead to decomposition and decay, and may increase the time that produce can be stored by several months. Ripening rooms, in which ethylene gas is added to the atmosphere, also help produce higher quality fruits and vegetables. This technology enables produce to be picked before it is ripe, for easier handling, and then ripened quickly and uniformly under controlled conditions. Other methods or a combination of methods may be used to preserve foods.

Modified atmosphere packaging (MAP)

In this method the air surrounding the food product is replaced by a different mixture of gases, where the proportion of each component is fixed when the mixture is introduced, but no further control is exercised during storage. Oxygen, Nitrogen and Carbon dioxide are the three major gases commercially used in MAP. Oxygen is important in fresh meats to maintain oxymyoglobin in it , which is responsible for giving fresh meat's bright red colour. Oxygen is not desirable in fatty fish, which may cause problems of oxidative rancidity or colour. Nitrogen is an inert tasteless gas with low solubility in both water and lipid. It is used to replace oxygen in packs to delay oxidative rancidity and inhibit the growth of aerobic microbes. It is used as filler gas to prevent the pack collapse , which may occur in high carbon dioxide atmosphere. Carbon dioxide is soluble both in water and lipid and is mainly responsible for bacteriostatic effects on aerobic microbes in MAP. MAP with Oxygen, Nitrogen and Carbon dioxide combinations suitable for each food has to be worked out based on moisture content, storage temperature etc.

Aseptic packaging

Aseptic packaging is now commonly used for packaging milk and juice. Like canning, aseptic packaging involves heat sterilization of food, but unlike canning, the package and food are sterilized separately. Food can be sterilized more rapidly and at lower temperatures in aseptic packaging than in canning, allowing the food to retain more nutrients and better flavour. Containers are sterilized with hydrogen peroxide rather than with heat, permitting the use of plastic bags and foil- lined cartons, which would be destroyed by heat sterilization. These containers cost less than the metal and glass containers used in canning and also weigh less, reducing transport costs. Aseptically packaged foods will keep without refrigeration for long periods of time, perhaps even years. They are growing in popularity because of their low cost, good taste and nutrition, and convenience.

Food package

Regardless of the processing or preservation method used, proper packaging of food is essential to make sure the food remains wholesome during its journey from processor to consumer. Package contains food and makes it easier to handle, and protects it from environmental conditions, such as temperature extremes, during transport. It locks out microorganisms and chemicals that could contaminate the food, and helps prevent physical and chemical changes and maintain the nutritional qualities of food. For example, milk is often stored in opaque containers to prevent vitamins from being destroyed by light.

Both the type of food and the processing method used affect the choice of packaging. For example, since oxygen makes fats go rancid, oils are packaged in containers that are impermeable to oxygen. On the other hand, oxygen-permeable plastic wraps allow fruits and vegetables to "breathe" and ensure that meats will maintain a vibrant red color. Metal and glass containers have traditionally been used in canning because these materials can withstand the high temperatures and changes in pressure that are involved in this processing method.

The development of metal cans in the early 1800s represented the birth of the modern packaging industry. The first British patent for a tin-plated steel container was issued in 1810 to British inventor Peter Durand. Canned foods were produced for the British armed forces in 1812 and offered commercially to the public two years later. Today, food cans are made of steel with various coatings to resist corrosion. Beverage cans are made of aluminum because it is lightweight and easy to manufacture. In addition to metal, glass is often used for packaging heat-sterilized foods. Glass is impermeable to oxygen and water and does not change the flavor of food. Another advantage of glass is that it is transparent, enabling the consumer to see the product inside. However, glass is not impact-resistant and is relatively heavy.

Plastic, by contrast, is lightweight and unbreakable, and it has become an extremely common material for use in food packaging. Most plastics used in food packaging are heat resistant so that they can go through high-temperature sterilization processes. Plastic is made into a wide variety of shapes, including bottles, jars, trays, and tubs, as well as thin films that are used as bags and wraps. By itself, paper is not frequently used in packaging, except for certain dry foods, such as flour and sugar. When paper is coated with plastic or other materials to make it stronger and impermeable to water, it can be more widely used. Paperboard is often used for cartons, and plastic-coated paperboard for packaging frozen foods. Cartons and containers for shipping are usually made of corrugated cardboard.

In recent years, environmental concerns have influenced food packaging. Scientists are working to develop packaging that is recyclable, biodegradable, or more compact so that it will use less landfill space, as well as to eliminate unnecessary packaging. Programs to recycle glass and aluminum beverage containers have been started all over the country. Plastic beverage bottles can be recycled as clothing or other products in addition to food containers. Aseptic packaging and several other new methods are compact and use a minimal amount of materials.

QUESTIONS

1. What are the principles of food preservation? Give critical account on various methods of food preservation by high temperature.

2. Describe the techniques involved in food preservation.
3. Compare the antimicrobial action of the following methods of food preservation
   i) Blanching
   ii) Canning
   iii) Pasteurization
   vi) Refrigeration.
4. Explain the various stages in canning of foods.
5. Define Pasteurization and its objectives. Explain different methods of Pasteurization of milk.
6. What are the objectives of blanching?
7. Give an account on various methods of freezing of foods and the lethal and sub-lethal effects of freezing on microorganisms.
8. Write a detailed account on irradiation of foods and doses prescribed for various foods. What are the advantages of irradiation?
9. Give an account on principles and methods the food irradiation.
10. List the characteristics of various radiations used in food preservation and explain the principle underlying in the destruction of microbes by irradiation.
11. List out various chemical preservatives used in food. What are the merits and demerits of using chemicals for food preservation?
12. What are the limitations of using chemical preservatives in foods?
13. Define: Food preservation, TDT, TDP, D value, Z value, Smoking, Pasteurization, Blanching, Food irradiation, Bacteriocin, Nisin.
14. A method of pasteurization that passes milk through a hot cylinder at 71.7°C for 15-17 seconds is called
    a) The holding method
    b) The flash method
    c) Ulltrapasteurization
    d) Sterile pasteurization
15. Which of the following is not a characteristic of gamma rays?
    a) Used to extend the shelf life of fruits, vegetables, and fish
    b) Increases the distance that fresh food can be transported
    c) Causes food to become radioactive at a low level
    d) Kills microbes by reacting with and destroying microbial DNA
16. Another name for freeze-drying is
    a) Irradiation
    b) Osmosis
    c) Lyophilization
    d) Smoking

3

FOOD SAFETY AND QUALITY MANAGEMENT

India is emerging as a food power in the international trade of foods. The immense un-tapped potential in the food processing sector in India has to be still exploited. It has been termed as 'sun rise industry' for improving agricultural productivity and enhancing food availability. The terms food quality management and food safety have assumed large dimensions today. In the international trade of foods, specifications have been laid down on the levels of harmful substances (chemical or biological origin), protocols have been documented on their detection and quantification in foods. We have to meet these international standards to export processed food and earn foreign exchange. The future of Indian food processing sector is bright provided it improves the quality of its products as per international standards.

3.1. FOOD SAFETY

The broad categories of food safety parameters of current interest are :

1. Internal additives (emulsifiers, stabilizers, leavening agents)
2. Indirect additives, toxicants and contaminants (pesticides)
3. Naturally occurring toxicants and food constituents (anti nutritional inhibitors)
4. Mycotoxins
5. Food borne bacterial intoxicants and infections
6. Food borne parasitic infections
7. Food and water associated viruses
8. Genetically modified foods.

In the 1960s the Pillsbury company, U.S army and NASA introduced a system for assuring pathogen-free foods for the space programme. This system called Hazard analysis Critical control point (HACCP) is focused on critical food safety areas as part of total quality programmes. The risk of the above mentioned safety parameters could be minimized by Hazard analysis Critical control point. It involves use of information about the factors leading to contamination, to establish preventive and control measures, thus providing the consumer with a safe food supply. Such HACCP programme could establish key risk elements to reduce the contaminant in field and storage environments. HACCP is basically a statement of controls based on the hazard analysis and critical control points.

Hazard analysis is aimed at identifying food hazards, biological, physical and chemical contaminants and maintains safety of foods. **Critical control point** is a point or step or procedure where control can be applied and hazard can be prevented, eliminated or reduced. The HACCP is a food safety system based on a set of safety regulations. The efficacy of the system depends essentially on proper identification of all relevant hazards associated with specific food under specific manufacturing, environment and processing parameters. Hence, proper guidance of relevant experts, especially the microbiologists, becomes important to the assurance of safety in processed foods to the consumer.

PHYSICAL HAZARDS: Glass, Wood, Stones, Metals, Bone, Plastic.

CHEMICAL HAZARDS: Allergens, Ciguatera toxin, Shellfish toxin, Mushroom toxin, Pesticides, fertilizers, antibiotics, heavy metals, plasticizers, paints, coding inks.

BIOLOGICAL HAZARDS IN FOODS

SPORE FORMING BACTERIA-*Clostridium botulinum, C. perfringens, Bacillus cereus.*

NON-SPORE FORMING BACTERIA- *Brucella sp. Campylobacter jejuni, Escherichia coli* 0157:H7, *Listeria monocytogenes, Salmonella* spp. *S. choleruesuis, S.enteritidis, Shigella dysenteriae, Staphylococcus aureus, Streptococcus pyogenes, Vibrio parahaemolyticus, Vibrio vulnificus, Yersinia enterocolitica.*

VIRUSES- Hepatitis A and E, Norwalk virus, Rota virus.

PROTOZOA - *Entamoeba histolytica, Ascaris lumbricoides.*

The preliminary steps in implementing HACCP system in food industries include, assembling the HACCP team, describe the food and its distribution, identifying intended use and consumers, developing and verifying flow diagram.

3.2. HACCP PRINCIPLES

There are seven principles in HACCP

1. Conduct hazard analysis
2. Identify critical control point
3. Establish critical limits for preventive measures.
4. Establish CCP monitoring system
5. Establish corrective action for deviation from CCP
6. Establish verification procedure
7. Establish record keeping procedure.

1. CONDUCT HAZARD ANALYSIS

Physical and chemical contaminants like environmental pollutants, food additives, pesticides and nutritional imbalance in foods are analyzed. The contaminants (hazards) are identified, strategies planned, implemented, monitored and reviewed or revalidated periodically. The microbial contaminants of foods are known to occur from a wide range of sources. In general there exists a close relationship between the source and type of contaminant leading to its predominance in the given environment. Foods may be contaminated by microorganisms at any time during harvest, storage, processing, distribution, handling, or preparation. The primary sources of microbial contamination are soil, air, animal feed, animal hides and intestines, plant surfaces, sewage, and food processing machinery or utensils. Since most foods either carry or eventually acquire bacteria, molds, or yeasts, microorganisms are the major causes of food spoilage. The increased load of microorganisms leads to rapid spoilage.

Preventive and control measures are planned based on the information about the factors leading to the contamination. The microbial growth is retarded by storage temperature, gases, additives, and natural antagonism and they are destroyed by various physical, chemical and combination of factors.

2. IDENTIFY CRITICAL CONTROL POINT

Determine the CCP to various steps/stages during food processing. During canning, the cans are filled, and then they are closed and sealed; a machine places the cover on the can, and the curl on the can cover and the flange on the can body are rolled into position and then flattened together. The thin layer of sealing compound originally present in the rim of the cover is dispersed between the layers of metal to ensure a hermetic seal. The sealed cans are then sterilized; *i.e.*, they are heated at temperatures high enough and for a long enough time to destroy all microorganisms (bacteria, molds, yeasts) that might still present in the food contents. The heating is done in high-pressure steam kettles, or cookers, usually using temperatures around 240°F (116°C).

During processing of canned foods, the hot cans (116°C) are immersed in cold water (30°C). During this step, water is sucked into the cans through very minute pores in the rim of the cover, leading to recontamination. The per cent of recontamination was found to depend on the microbiological quality of water used for cooling the cans. The per cent of re-contaminated cans increased with increase in the bacterial population in cooling water (Table 3.1).

TABLE 3.1 CRITICAL CONTROL POINT IN RE-CONTAMINATION OF CANNED FOODS

BACTERIAL POPULATION IN WATER PER ML	PER CENT OF CANS RE-CONTAMINATED
10^2	2
10^3	8
10^4	18
10^5	30
10^6	48
10^7	62

Chlorination of water was the control applied to reduce bacterial population in cooling water. The per cent of re-contaminated cans was reduced by chlorination in canning industry.

The breads, cakes, sweet rolls, and some other bakery foods may contain as much as 38 to 40 percent water when freshly baked and are subject to attack by many fungi and a few bacteria. To obtain maximum shelf life free of mold spoilage, high levels of sanitation must be maintained in baking and packing areas. Oven heat destroys all fungal life-forms, and any spoilage by these organisms is due to re-inoculation after baking. In a bakery industry, the mold contamination and infestation were high. When examined it was found that the contamination came from mold infested dampened walls of the bakery. To make the place mold free wall painting in the bakery was the control applied.

The most common CCP is cooking, where food safety managers designate critical limits. The minimum internal temperatures for cooked foods are given below (Table 3.2).

TABLE 3.2 CRITICAL CONTROL POINT
Minimum internal temperatures for cooked foods

FOOD	TEMPERATURE AND TIME
Cooked poultry, meat, fish or eggs	74 °C for 15 seconds
Ground meat, ground or minced fish	68 °C for 15 seconds
Cooked fruit or vegetables	57 °C for 15 seconds

In addition, hot food must be held at a minimum internal temperature of 57° C if it is not immediately consumed.

3. ESTABLISH CRITICAL LIMITS FOR PREVENTIVE MEASURES

Establish critical limits for preventive measures during the various stages of food processing. The flow diagram of the processing unit with the stages of processes and number of hours is prepared for material and time study. In processing of canned vegetables, cleaning, blanching, brining and packing are the various stages. Temperature, pH, sodium chloride concentration of solution, are the critical limits in vegetables processing. The pH of many vegetables is neutral and they are acidified to prevent microbial contamination.

The temperature employed for blanching is decided based on the heat resistance of microbial contaminants expected in the vegetable. The sodium chloride concentration of brine solution is selected based on the product and the salt tolerance of microbial contaminants expected in the vegetable. The microbial contaminant populations at various levels of these critical limits - pH, temperature, sodium chloride concentration, are studied and preventive measures are planned.

4. ESTABLISH CCP MONITORING SYSTEM

Establish a system to monitor control of the CCP. The quality control system procedure should be followed in spirit and not in letters. The quality control unit should function independently free from interference. All quality parameters should be tested and recorded for each batch of the product. A multi-disciplinary team comprising plant engineer, microbiologist, quality manager, market manager, consultant, auditor and General Manager should be given training on quality control. The workers should be quality conscious and given periodical training, as ignorance is the reason for disobedience. Inter and intra group communication is maintained in the processing industry to ensure there is no gap. Customer's feedback and complaints are addressed to improve the quality.

5. ESTABLISH CORRECTIVE ACTION FOR DEVIATION FROM CCP

Establish the corrective action to be taken when monitoring indicates that a particular CCP is not under control. Market intelligence team should monitor the consumer response and defective product is recalled. In a fruit juice industry, the fruit juice was bitter in taste as it was contaminated with a water soluble toxin. The corrective action taken was, the juice mixed with water to solubilize the toxin and water with toxin filtered to remove the toxin. The team has to find out what went wrong, learn from the experience and do not let it happen again.

6. ESTABLISH VERIFICATION PROCEDURE

Establish procedures for verification to confirm that the HACCP system is working effectively. The top management should have commitment for quality in all aspects. Whether all the quality assurance procedures are followed is verified. Quality assurance focuses more on prevention than cure. The quality assurance procedures are integrated and harmonized into one system. An external agency, like Bureau of Indian Standards (BIS) or International Standard Organization (ISO) should evaluate the quality of the unit. ISO 9000 is a fully documented quality management system. ISO 9001 covers design and development in R&D organizations; ISO 9002 covers production unit; ISO 9003 covers installation and servicing organizations; ISO 14000 series is for environmental management.

7. ESTABLISH RECORD KEEPING PROCEDURE

Establish documentation concerning all procedures and records appropriate to these principles and their application. The policy, vision and mission of the unit should be clearly displayed in letters. Documentation of all the procedures in the processing industry is important for improvement. The results of all the activities of the unit are recorded. Record keeping helps to solve the problems encountered earlier. The activities of the unit should not be person oriented, but record oriented.

Reliable means to assure food safety and protect consumer's health are the goals of HACCP concept based in plant control programs rather than sporadic microbiological monitoring of end products.

3.3 FOOD STANDARDS

For hundreds of years, governments have had an interest in regulating food processing to ensure the safety and wholesomeness of the foods consumed by their citizens. The earliest known food law was written in Japan in AD 702. In Britain, the first Pure Food Laws were enacted during the 1860s to combat adulteration, the secret use of additives to stretch wholesome foods with cheaper, non-nutritious (and sometimes dangerous) ingredients. This practice became common during the Middle Ages (5th century to 15th century), when cities began to grow and urban populations no longer got their food directly from the farm, creating an opportunity for deception by middlemen. Today, several United States government agencies carry out inspections and enforce a comprehensive system of regulations governing food processing, packaging, and distribution.

Among the first regulations in the United States for protecting the food supply was the Meat Inspection Act of 1906. It was inspired in part by the 1906 novel, The Jungle by Upton Sinclair, which described unsanitary conditions in Chicago stockyards. The Meat Inspection Act required inspection of cattle, hogs, sheep, and goats intended for interstate commerce, as well as the monitoring of slaughter and processing procedures. When poultry became more popular during the 1950s, a similar law, the Poultry Products Inspection Act of 1957, subjected poultry producers to the same scrutiny.

Today the Food Safety and Inspection Service, a part of the Department of Agriculture (USDA), carries out inspections of meat and poultry processing. Animals are inspected by USDA veterinarians before they are slaughtered to eliminate animals with disease from the food supply. This phase of inspection includes blood tests of randomly selected animals from each flock to make sure residues of antibiotics or pesticides, if present in the animals' blood, do not exceed acceptable levels. Animals are also inspected after slaughter for disease or contamination of carcasses, and processing procedures, recipes, and labels on products are also monitored.

The FDA is responsible for monitoring and inspecting most all other food products. The FDA enforces the Food, Drug, and Cosmetic Act of 1938, which prohibits the shipping of adulterated or mislabeled products in interstate commerce. FDA inspectors visit food processing plants and warehouses to monitor all phases of processing, packaging, and distribution. Samples of food products are analyzed by FDA chemists to ensure the foods are wholesome and unadulterated and do not contain harmful substances, such as levels of pesticides above the limits set by the Environmental Protection Agency (EPA). In 1958, amendments to the Food, Drug, and Cosmetic Act gave the FDA authority to regulate the use of additives in foods. The Delaney Clause, which was part of those amendments, prohibits approval of additives that cause cancer in human or animal tests. Some people have criticized this clause for being too inflexible, because some studies show very small increases in cancer risk and some studies that show a risk of cancer in animals may not be relevant to humans. Since 1969, the FDA has set standards for inspection of retail food stores, restaurants, and cafeterias, although local health departments are responsible for carrying out those inspections.

Package labels are also regulated by the federal government. What began as an effort to make sure that labels were accurate has now expanded to require labels that provide more information to increasingly health-conscious consumers. The 1938 Food, Drug, and Cosmetic Act prohibited mislabeling of food. In 1964 the Truth in Labeling Act required the package label to provide an accurate description of the weight, volume, or count of the package contents, a description of the contents, and the name and address of the manufacturer. The Nutrition Labeling and Education Act of 1990 required that all packages include a label, headed "Nutrition Facts." This label must list the serving size and number of servings per package, and detail the food's content of various nutrients (including carbohydrates, proteins, fats, vitamins, and minerals). These labels help people compare different food choices and select those that fit their diet.

The Codex Alimentarius Commission (**Fig. 3.1**) was created in 1963 by FAO and WHO to develop food standards, guidelines and related texts such as codes of practice under the Joint FAO/WHO Food Standards Programme. 'Codex Alimentarius' means - code of food standards. The Codex Alimentarius is a collection of international food standards covering all the main foods and also material used in the further processing of food. The main purposes of this Programme are protecting health of the consumers and ensuring fair trade practices in the food trade, and promoting coordination of all food standards work undertaken by international governmental and non-governmental organizations. There are 180 codex members in the world. It initiates and guides the preparation of draft standards through appropriate organizations, finalize standards and after acceptance by Governments publish them in a Codex Alimentarius. Codex has produced 237 food commodity standards, 41 hygienic and technological practice codex, evaluated 785 food additives and contaminants and has set maximum residue limits for 3200 pesticides (**Table 3.3**).

TABLE 3.3 LIST OF STANDARDS AND OTHER RELATED TEXT ADOPTED BY THE CODEX ALIMENTARIUS COMMISSION

VOLUME	REFERENCE	TITLE
Vol 01 A	**CAC/RCP-020-1985** (Recommended code of practice)	Code of ethics for international trade in food
Vol 01 A	**CAC/RCP -045 -1997**	Code of practice for the reduction of Aflatoxin in raw materials and feed stuff for milk producing animals
Vol 01 A	**CODEX-STAN -209-1999** (STANDARD)	Standard for maximum level for aflatoxin in pea nuts intended for further processing.
Vol 01 B	**CAC/ GL -021 -1997** (Guide lines)	Principles for the establishment and application of microbiological criteria for food
Vol 01 B	**CAC/ GL -030 - 1999**	Principles and guide lines for the conduct of microbiological risk assessment
Vol 04	**CAC/RCP- 021-1979**	Recommended International code of hygienic practice for foods for infants and children (including microbiological specifications and methods for microbiological analysis)
Vol 05 A	**CAC/RCP - 005-1971**	Recommended international code of hygienic practice for dehydrated fruits and vegetables including edible fungi
Vol 05 A	**CODEX-STAN 041-1981** (STANDARD)	Standard for quick frozen Peas
Vol 05 A	**CODEX-STAN -042-1981** (STANDARD)	Standard for canned Pineapple
Vol 06	**CODEX/STAN 139-1983**	Standard for concentrated pine apple juice with preservatives for manufacturing
Vol 05 A	**CODEX-STAN 055-1981** (STANDARD)	Standard for canned Mushrooms

VOLUME	REFERENCE	TITLE
Vol 05 A	**CODEX-STAN 115-1981** (STANDARD)	Standard for Pickled cucumbers
Vol 05 A	**CODEX-STAN 143-1985** (STANDARD)	Standard for Dates
Vol 05 B	**CODEX-STAN 183-1993** (STANDARD)	Standard for Papaya
Vol 07	**Codex STAN 152 -1985**	Standard for wheat flour
Vol 05 B	**CODEX-STAN 184-1993** (STANDARD)	Standard for Mango
Vol 05 A	**CODEX-STAN 174-1989** (STANDARD)	General Standards for Vegetable Protein Products
Vol 07	**Codex STAN 152 -1985**	Standard for wheat flour
Vol 10	**CAC/ GL 014 -1991** (Guide lines)	Guide for microbiological Quality of Spices and Herbs used in processed meat and poultry products
Vol 11	**CODEX-STAN 108-1981** (STANDARD)	Standards for Natural Mineral Water

Fig. 3.1 Codex Alimentarius Commission logo

The standards contain requirements for food aimed at ensuring the consumer a sound, wholesome food product free from adulteration, correctly labeled and presented. The people everywhere have the basic right of access to food which is of good quality, safe and nutritious. Codex Alimentarius helps to reach this goal by combining consumer protection with food production and trade, globally, regionally and at national and local level.

The categories covered in the 237 codex commodity food standards are :

1. Processed fruits and vegetables and edible fungi
2. Sugars
3. Processed meat, poultry, products
4. Fish and fishery products
5. Cocoa products and chocolates
6. Quick frozen fruits and vegetables
7. Concentrated fruit juices and fruit nectars
8. Edible fats and oils
9. Milk products
10. Cereals, pulses, legumes and derived products
11. Edible ices and ice mixes
12. Vegetable proteins.

According to Codex, a food label has to contain - (1) name of the food, (2) list of ingredients, (3) net contents and drained weight, (4) name and address of the manufacturer, (5) country of origin, (6) date of making and storage instructions, (7) instructions for use.

3.4. FOOD SAFETY IN INDIA

In India, the Prevention of Food Adulteration (PFA) Act, 1954 is the major law to protect the consumer against adulterated foods. The whole range of requirements are covered under the PFA Act, which includes sampling and analysis of foods, food additives and their limits, labeling requirements, food inspection and conditions for sale and license. The Bureau of Indian Standards (BIS) and Directorate of Agricultural Marketing and Inspection (AGMARK) are the organizations working for voluntary standards and collaborating with International Standard Organization (ISO). The recent strategy is to instruct food processing industries to obtain ISO certification by following the guide lines of ISO regarding procurement, pretreatment, manufacturing practices, quality control, food standards, documentation, etc. The national and international organizations work on a parallel platform, to cite an example, ISO 9000 series a fully documented quality management system is equivalent to BIS140000 series. Thus Total Quality Management (TQM) system, which is a combination, coordination and compliance of all safety and quality control measures with quality assurance programmes are to be followed by food processing industries.

Bureau of Indian Standard (BIS) adopted the Indian standard for microbiological quality of foodstuffs finalized by the food hygiene, sampling and analysis sectional committee and approved by the Agriculture and food products division council of Government of India. The Indian standard, IS-5401-1969 and 1991-METHOD FOR DETECTION AND ESTIMATION OF COLIFORM BACTERIA IN FOOD STUFFS, specifies the sampling, apparatus and materials, media and procedure for detection and estimation of coliform bacteria in foodstuffs. Interpretation of results: presence of more than 10 coliform colonies per gram (positive test in 0.1 ml of liquid media) in the case of heat processed foods and of presence of more than 100 coliform colonies per gram (positive test in 0.01 ml of liquid media) in the case of un-processed materials indicates insanitary methods of production and handling.

The Food Safety and Standards Authority of India (FSSAI) has been established under Food Safety and Standards Act, 2006 which consolidates various acts and orders that have hitherto handled food related issues in various Ministries and Departments. FSSAI has been created for laying down science based standards for articles of food and to regulate their manufacture, storage, distribution, sale and import to ensure availability of safe and wholesome food for human consumption.

HIGHLIGHTS OF THE FOOD SAFETY AND STANDARD ACT, 2006

Various central Acts like Prevention of Food Adulteration Act 1954, Fruit Products Order 1955, Meat Food Products Order 1973,Vegetable Oil Products (Control) Order 1947, Edible Oils Packaging (Regulation) Order 1988, Solvent Extracted Oil, De-Oiled Meal and Edible Flour (Control) Order 1967, Milk and Milk Products Order 1992, etc will be repealed after commencement of FSS Act 2006.

The Act also aims to establish a single reference point for all matters relating to food safety and standards, by moving from multi- level, multi-departmental control to a single line of command. To this effect, the Act establishes an independent statutory Authority - the Food Safety and Standards Authority of India with head office at Delhi. Food Safety and Standards Authority of India (FSSAI) and the State Food Safety Authorities shall enforce various provisions of the Act. The Important Food Acts and Orders of India (Table 3.4).

TABLE 3.4 THE IMPORTANT FOOD ACTS AND ORDERS OF INDIA

FOOD ACT	AGENCY
1. The Prevention of Food Adulteration Act 1954	Directorate General of Health services, New Delhi, under Ministry of Health and Family welfare
2. AGMARK (Agri. Products Marketing) Act 1937	Department of Marketing and Inspection, Ministry of Agriculture
3. Sugar control Order, 1965	Ministry of Agriculture and Irrigation.
4. Fruit Product Order, 1955	Department of Food Processing Industries, Ministry of Agriculture
5. Vegetable Oil Products (control) Order, 1947	Department of Vanaspati, Ministry of Civil Supplies and Co-operation.
6. Export Quality control and Inspection Act, 1963 and Rules 1964	Ministry of Commerce
7. Meat Food Products Order, 1975	Department of Marketing and Inspection, Ministry of Civil Supplies and Co-operation
8. The consumer Protection Act, 1986.	Ministry of consumer affairs, food and public distribution
9. Indian Standard Institution-Bureau of Indian Standard	Ministry of Industries and Civil Supplies.
10. Food Safety and Standards Act, 2006	Food Safety and Standards Authority of India (FSSAI)

FSSAI has been mandated by the FSS Act, 2006 for performing the following functions

- Framing of Regulations to lay down the Standards and guidelines in relation to articles of food and specifying appropriate system of enforcing various standards thus notified.
- Laying down mechanisms and guidelines for accreditation of certification bodies engaged in certification of food safety management system for food businesses.
- Laying down procedure and guidelines for accreditation of laboratories and notification of the accredited laboratories.
- To provide scientific advice and technical support to Central Government and State Governments in the matters of framing the policy and rules in areas which have a direct or indirect bearing of food safety and nutrition.

- Collect and collate data regarding food consumption, incidence and prevalence of biological risk, contaminants in food, residues of various, contaminants in foods products, identification of emerging risks and introduction of rapid alert system.
- Creating an information network across the country so that the public, consumers, Panchayats etc receive rapid, reliable and objective information about food safety and issues of concern.
- Provide training programmes for persons who are involved or intend to get involved in food businesses.
- Contribute to the development of international technical standards for food, sanitary and phyto-sanitary standards.
- Promote general awareness about food safety and food standards.

3.5. FOOD SAFETY MANAGEMENT SYSTEM (FSMS) FOR SAFE FOOD SUPPLY

Food Safety is related to the presence of food borne hazards in food at the point of consumption. Food reaches to consumers via supply chains that may link many different types of organizations. One weak link can result in unsafe food that is dangerous to health. As food safety hazards can occur in the food chain at any stage, adequate control throughout the supply chain is essential. Therefore food safety is a joint responsibility of all organizations with in the food chain including, producers, manufactures, transport and storage operators, sub contractors, retail and food service outlets and service providers.

Recent studies have shown that there is significant increase of illness caused by infected food in both developed and developing countries which give rise to considerable economic costs besides being health hazards. This has necessitated the need for establishing a food safety management system by all types of organizations within the food chain.

HIGHLIGHTS OF IS/ISO 22000:2005

a) Integrates the principles of Hazards Analysis and Critical Control Point (HACCP) system developed by Codex Alimentarius Commission. It combines the HACCP plan with prerequisite programmes (PRPs) and operational PRPs.

b) Requires that all Hazards that may be reasonably expected to occur in the food chain are identified, assessed and controlled.

c) Can be applied independent of other management system standards or can be integrated with existing other management systems.

d) Allows even small, tiny scale organizations to implement as externally developed combination of control measures.

e) Intended for organizations seeking more focused, coherent and integrated food safety management systems.

f) Emphasis on preventions of food safety hazards of all types.

g) Ensures compliance with legislative and regulatory requirements.

h) Provides for management of potential emergency situations and accidents that can impact food safety.

ISO22000 series of standards for food safety management system are followed in 60 countries by 200 organizations. The key elements of FSMS for managing and reducing the risk to health resulting from operations across the food chain to final consumption are: Interactive Communication; System Management; Prerequisite Programmes; HACCP Principles.

The benefits to customers are: Increased international acceptance of food products; Reduces risk of product/service liability claims; Satisfies customer contractual requirements; Ensures safety of food products; Greater health protection; Demonstrations conformance to international standards and applicable regulatory requirements; Helps to meet applicable food safety related statuary and regulatory requirements; Ensures to compete effectively in national and international markets. FAO's Food Quality and Standards Service is committed to the enhancement of food safety and quality along the food chain at international, regional and national levels, with the aim of protecting consumers and promoting the production and trade of safe, quality food.

Good manufacturing practices (GMP) is defined as a set current scientifically sound methods/ practices/ principles, that are implemented and documented during food production to ensure consistent manufacture of safe, pure and potent products. GMP should be followed in the manufacture, processing and storage of foods.GMP is aimed at quality assurance and a tool in microbial environment monitoring. Clean room, design, operation layout, air handling system, utilities, operator qualifications are important aspects of GMP in manufacture of products. To achieve highest level of safety, purity and efficacy of products, people are moving beyond quality control (QC) and into quality assurance (QA).

World Food Day is celebrated every year on 16 October to commemorate the founding of the Food and Agriculture Organization (FAO) of the United Nations in 1945. World Food Day aims to heighten public awareness of the plight of the world's hungry and malnourished and to encourage people worldwide to take action against hunger. More than 150 countries observe this event every year.

QUESTIONS

1. What are the broad categories of food safety parameters of current interest?
2. What is HACCP? Explain the seven principles of HACCP.
3. Mention the physical, chemical and biological hazards in foods.
4. Write an account on food standards.
5. Discuss the role of Codex Alimentarius commission in food quality control.
6. Explain the role of Indian agencies in food safety and quality control.
7. What are the important food acts and orders in force in India?
8. Codex alimentarius commission is a joint organization of

   (a) FAO & WHO (b) ILO & WTO

   (c) UNESCO & UNO (d) IMF & ISI
9. The principles of HACCP are integrated in

   (a) ISO 22000 (b) ISO 14000

   (c) ISO 9000 (d) ISO 9001

4

FERMENTED FOOD PRODUCTS

Fermentation plays a major role in our diet and give the basic foods like bread, yoghurt, wine and cereal and legume based traditional fermented foods. The development of fermented food products can be considered as one of the greatest achievements of human civilization and goes back thousands of years. It arose from the necessity to conserve foods, to make them more digestible and also more enjoyable. During the centuries, man learned by trial and error to conduct the fermentation processes by changing physical and chemical parameters in the basic food ingredients. It took much longer, however, to discover the cause of fermentation processes: only towards the end of the 19th century it was demonstrated, that baker's yeast is the cause of alcoholic fermentation. Through the development of the technique of "pure microbial culture", the basis for a controlled use of microbes in food industry was established.

The unique feature of fermented foods is the vital role of microorganisms that bring about essential bio-trasformations of the substrates during fermentation contributing number of desirable properties such as improved product shelf life, enriched diet with improved flavor and texture, increased safety, enriched nutritional supplement and probiotic functions. Fermentation is a chemical reaction carried out by many types of microorganisms to obtain energy. Fermentation is a process for production of a product by means of mass culture of microorganisms. In fermentation, microorganisms breakdown complex organic compounds into simpler substances. Although chemical changes and microbial growth usually mean food spoilage, in some cases fermentation is desirable and microorganisms are actually added to foods. For example, in the production of beer, wine, and other alcoholic beverages, yeasts convert sugar into ethyl alcohol and carbon dioxide. In the making of yogurt and cheese, lactic acid bacteria convert lactose, a sugar found in milk, to lactic acid. Alcohol, acids, and other compounds produced in fermentation act as preservatives, inhibiting further microbial growth. In addition to its use with alcoholic beverages, cheese, and yoghurt, fermentation is used to produce yeast bread, soy sauce, cucumber pickles, sauerkraut, and other traditional fermented food products. The variation in agro-climatic conditions prevailing in India and diverse form of dietary culture of various ethnical groups produced traditional fermented foods such as idli, dosa (rice-legume based), vada, papad, wari, balle (legume based), bhatura, jelabi, nan (wheat based), paneer, shrikhand (milk based).

Yeasts can be considered man's oldest industrial microorganism. Ancient Egyptians were using yeast and the process of fermentation to produce alcoholic beverages and to leaven bread over 5000 years ago. The biochemical process of fermentation that is responsible for this action was not understood. These early fermentations of alcohol production and bread making were formed by natural microbial contaminants of the raw materials. Such microbial flora would have included wild yeasts and lactic acid bacteria. The pioneering scientific work of Louis Pasteur in late 1860s revealed that yeast as a living organism and the agent responsible for alcoholic fermentation and dough leavening. Yeasts are single celled fungi and over 600 species of yeasts are known to be distributed in nature. They are normal inhabitants of soil, vegetation, marine and other aqueous environments.

Fermented foods are defined as food that has been subjected to the action of selected microorganisms by which a biochemically and organoleptically modified substrate is produced (Tamang, 1998). Distribution of the numerous fermented foods in the world and their manufacturing processes are determined by the kinds of raw materials, principal microbes, environmental conditions and food habits found in each region. Fermented foods can be divided in to nine classes: i.Beverages (wine), ii.Cereal products (bread), iii.Dairy products (yoghurt), iv. Fish products (fish sauce), v. Fruit and vegetable products (sausages), vi. Legumes (soya sauce), vii. Meat products, viii. Starch crop products, and ix. Miscellaneous products. Many of these classes of foods are specific to particular regions of the World and in many cases, they are important factors in local diet.

The microbes used for (or) found in food fermentations are also conveniently divided as follows: a) Yeast, b) Bacteria, c) Filamentous fungi. Yeasts are in the genera, *Saccharomyces, Zygosaccharomyces, Entomycopsis* and *Torulopsis,* Fermenting bacteria include the genera *Acetobacter, Lactobacillus, Pediococcus, Micrococcus, Clostridium* and

Bacillus, and the filamentous fungi include the genera *Aspergillus*, *Penicillium*, *Rhizopus* and *Mucor*.

According to Steinkraus (1995), food fermentation serves five main purposes.

- Enrichment of the diet through development of a diversity of flavours, aromas, and textures in food substrates.
- Preservation of substantial amounts of food through lactic acid, alcoholic, acetic acid, and alkaline fermentations.
- Biological enrichment of food substrates with protein, essential amino acids, essential fatty acids, vitamins (B group vitamins show an increase on fermentation) and improves the bioavailability of minerals.
- Detoxification during food-fermentation processing and removal of anti-nutrients.
- A decrease in cooking times and fuel requirements.

Fermentation has some uses exclusive to foods. Fermentation can produce important nutrients or eliminate anti-nutrients. Fermentation is the most economical process for preservation of the quality and safety of foods. Food can be preserved by fermentation, since fermentation uses up food energy and can make conditions unsuitable for undesirable microorganisms. For example, in pickling the acid produced by the dominant bacteria inhibit the growth of all other microorganisms.

4.1. YEAST FERMENTATION

When yeast ferments, it breaks down the glucose ($C_6H_{12}O_6$) into exactly two molecules of ethanol (C_2H_6O) and two molecules of carbon dioxide (CO_2).

$$C_6H_{12}O_6 \rightarrow 2\ CO_2 + 2C_2H_5OH.$$

The primary benefit of fermentation is the conversion, e.g., converting juice into wine, grains into beer, and carbohydrates into carbon dioxide to leaven bread.

4.1.1. Wine

The practice of making wine is as old as our most ancient civilization. Archaeological evidence suggests that wine was being made at least 8,000 years ago, and there are indications that viticulture was carried on in Mesopotamia (the region that is now Iraq and eastern Syria) about 5,000 years ago. Writings from the same time from China and Egypt also mention wine and vineyards. Egyptian frescoes show festive grape harvest scenes, an atmosphere that still prevails in wine-producing countries during the harvest season. By the time of early Greek civilization, about 3,500 years ago, wine was a popular beverage as well as a sacred drink. In Greek mythology, Dionysus was the god of wine, overseeing all matters pertaining to the cultivation of grapes and wine production. The ancient Greeks were the first to cultivate grapes for commercial purposes, and to market their wine abroad. Wine was equally important to the ancient Romans who looked to Bacchus, the Roman god of wine, to oversee all wine matters. In the highly organized economy of the Roman empire, viticulture was centered in the areas of present-day Italy, Spain, Greece, France, and Germany. Some of the finest vineyards in France and Germany have been under cultivation since this time.

Wine is a product of alcoholic fermentation of fruit juice that is rich in fermentable sugars. During fermentation the yeast digest sugar found in fruit juice, producing alcohol and carbon dioxide in the process. Most wines are made from grapes only. Wine naturally contains 85-89% water, 10-14% alcohol, < 1% fruit acids and hundreds of flavour and aroma compounds in very small amounts.

TYPES: There are many types of wine and three basic groups of wines are most easily distinguishable for the consumer: table wines, fortified wines and sparkling wines. Table wines, also known as still or natural wines, are produced in many different styles and make up the majority of wines on the market. Traditionally consumed as part of a meal, table wines contain between 10 and 14 percent alcohol and are further classified by their colour, sugar content, and the variety and origin of the grapes that were used. Depending on the grape variety and wine-making technique, wines can be white, red, or pink in colour. Red wines are prepared by fermentation of grape juice along with the grape skins. Red pigments called anthocyanins and other compounds in the grape skins are extracted during the fermentation process to impart the characteristic red color of the wine as well as other features. White wine, which is actually straw to golden-yellow in color, is produced from white grapes, by taking the grape juice only. White grapes are crushed and the juice separated from the skins prior to fermentation. A blush or rosé wine is light pink in color and is produced from red grapes not fermented with the skins. A little pigment is released when the red grapes are crushed, but not to the same extent as during fermentation with the skins.

Most table wines are fermented until they are dry-that is, all the grape sugar has been turned to alcohol by the yeast. In dry wines all the sugars are perfectly fermented whereas in sweet wines some of the sugars are left unfermented or additional sugar is added after fermentation. In fortified wines brandy or alcoholic spirit is added after fermentation. Fortified wines contain additional alcohol and are usually consumed in small amounts as aperitif before meals or dessert wines after a meal. Popular examples are port and sherry. In port wine making, which originated in Portugal, the grapes are crushed and the fermentation started but then stopped by the addition of more alcohol, which kills the yeast. The resulting wine is sweet and has an alcohol content that is 5 to 10 percent higher than table wine. Most of the wines have 10 -12 % alcohol whereas the dessert wines contain 20% alcohol. Sparkling wine is made from table wine that has undergone a second fermentation. The wine maker adds a measured amount of sugar and fresh yeast to the dry wine. This can happen in a closed tank, or directly in the bottle, which is the way the most famous sparkling wine, French champagne, is produced. The yeast ferments the added sugar, but this time the carbon dioxide gas remains in the sealed bottle, creating carbonation. When the sparkling wine is poured into a glass, the gas bubbles are released to the surface. In sparkling wine considerable carbon dioxide is present arising from final fermentation.

RIPENING: Fresh and fully ripened wine grapes are preferred as raw material for wine making. In cool climates, as in northern Europe and the eastern United States, however, lack of sufficient heat to produce ripening may necessitate harvesting the grapes before they reach full maturity. The resulting sugar deficiency may be corrected by direct addition of sugar or by the addition of a grape juice concentrate. Grapes that are allowed to reach full maturity on the vine or that are partially dried by exposure to sun after harvesting are high in sugar content as a result of natural moisture loss (raisin). A beneficent mold, *Botrytis cinerea,* may also be employed to hasten moisture loss. These grapes are used to produce sweet table wines. Special methods employed to produce these wines include the addition of sulfur dioxide, the use of small fermenting vessels during processing, or the use of cool temperatures- the objective being to stop the fermentation before all the sugar is fermented.

HARVEST: Because of the effect upon grape composition, proper timing of the harvest is of great importance. Premature harvesting results in thin, low-alcohol wines; very late harvesting may yield high-alcohol, low-acid wines. Harvesting may be completed in one picking or in several. The grape clusters are cut from the vine and placed in buckets or boxes and then transferred to larger containers (large tubs in Europe, metal gondola trucks in California and elsewhere) for transport to the winery. Mechanical harvesting systems, based on shaking the berries from the clusters or on breaking the stems, are widely used in California, Australia, France, and elsewhere. At the winery the grapes may be dumped directly into the crusher or may be unloaded into a sump and carried to the crusher by a continuous conveyor system.

Wine Making

Red grapes are sometimes introduced whole into tanks, which are then closed. The resulting respiration in the fruit, consuming oxygen and producing carbon dioxide, kills the skin cells, which lose their semi-permeability, allowing easy colour extraction. There is also some intracellular respiration of malic acid. This respiration process is slow and in warm regions may result in wines of low colour and acidity and distinctive odour.

In modern mechanized wine production, the grapes are normally crushed and stemmed at the same time by a crusher- stemmer, usually consisting of a perforated cylinder containing paddles revolving at 600 to 1,200 revolutions per minute. The grape berries are crushed and fall through the cylinder perforations; most of the stems pass out of the end of the cylinder. A roller-crusher may also be used. Ancient methods of crushing with the feet or treading with shoes are rare. When red grapes are used to produce a white juice, as in the Champagne region of France, crushing is accomplished by pressing.

Stages in Wine Making

Fully ripened grapes with sweet taste and optimum flavour are selected and the stem and stalks of the berries are removed as they contain tannin. Grapes are washed with clean water and crushed to yield grape juice. The crushed juice called Must has a pH of 3.0. If the sweetness of Must is low (Brix value < 20), sugar is added. SO_2 is added to remove wild yeast contaminants and potassium meta bisulphite is added at the rate of 1.0 mg per kg of grapes as a preservative. After six hours of adding SO_2 the starter culture is added. The Must is inoculated with 3% yeast starter culture, *Saccharomyces cerevisiae* var *ellipsoideus*. Nutrients like, ammonium sulphate is added as nitrogen source. Anaerobic condition, a temperature of 21-27°C and pH at 3.0-3.6 are maintained during fermentation.

Partially fermented wine is decanted and again allowed to ferment for 7-10 days. After fermentation is completed the wine is decanted from sediment of yeast and particles of grape tissues. Fermented Must is filtered and fining agents such as casein, gelatin, egg white, and tannin may be added for clarification. Clearing of wine and development of flavour take place during aging of at least 8 - 10 months. The aged wine is again filtered, pasteurized and bottled before consumption.

FLOW CHART 4.1. WINE MAKING

BIOCHEMICAL REACTION

$C_6H_{12}O_6 \rightarrow 2\ CO_2 + 2C_2H_5OH$

GRAPES PICKING, REMOVAL OF stem, stalks etc.

↓

WASHING IN CLEAN WATER

↓

WEIGHING, CRUSHING

↓

MUST -Brix value < 20 ° need addition of sugar

↓

SO_2 ADDED to destroy wild yeast

↓

YEAST INOCULUM at 3 %

Saccharomyces cerevisiae var *ellipsoideus*

↓

FERMENTATION

↓

DECANTING

Fermented material passed through a press, Skin and seeds removed

↓

CLARIFICATION

Fining material added

↓

AGING OF WINE : 8-10 months

↓

FILTRATION, PASTEURIZATION

↓

BOTTLED/ PACKED

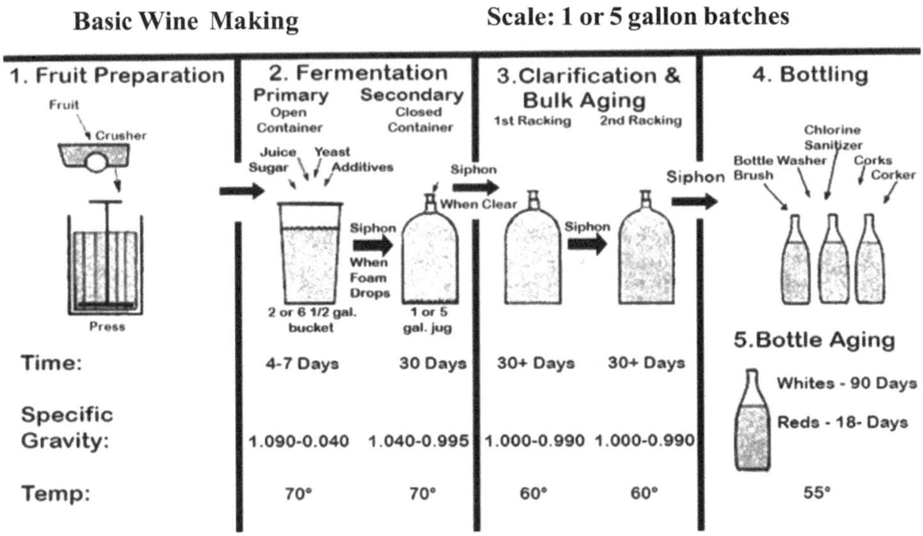

Fig. 4.1 Basic Wine making

Aging of Wine

Many wines improve in quality during barrel and bottle storage. Such wines eventually reach their peak and with further aging begin to decline. During the aging period, acidity decreases, additional clarification and stabilization occur as undesirable substances are precipitated, and the various components of the wine form complex compounds affecting flavour and aroma. Wines are usually aged in wooden containers made of oak, allowing oxygen to enter and water and alcohol to escape. Extracts from the wood contribute to flavour.

Some red table wines appreciate in quality, developing less astringency and colour, and a greater complexity of flavour with aging in oak cooperage of up to 500-gallon size for two to three years. In the best red wines, additional improvement may continue with two to 20 years of bottle aging (the rate of aging being lower in the bottle than in the barrel). Many dessert wines improve during cask aging, particularly sweet cherries, but extraction of excessive wood flavour must be avoided. Those rosé and dry red wines that will not improve with long cask and bottle aging are aged for a short period of time, clarified, and then bottled. More than 90 percent of all table wines are probably marketed and consumed before they are two years old. The chief benefit of aging is greater clarification as various undesirable substances are precipitated. Wine matures during storage, which generally varies between 3 and 9 months, but which may be extended up to three years for high quality wines. The maturation or aging is an important step in the manufacture of wine that produces desirable changes in body and flavour of wine. During maturation some chemical reactions takes place and bouquet, the flavour of wine is developed through oxidation and esterification. The wine is allowed for aging in new oak casks or bottles. Oak wood has several complex chemical compounds; each of it contributes its own flavour or textural note to wine. The colour also changes during aging from purple to carmine in red wine and yellow to gold in white wine. The age of the wine generally affects the flavour and with experience the taster can tell much about the age of the wine by its flavour. Wine looses some of its astringency as the tannin settle and acids are converted into ether, resulting in a characteristic 'bouquet'.

The characteristic flavour develops due to various volatile aroma substances produced in wine during maturation. These substances include esters (ethyl formate, methyl acetate, and ethyl acetate), aldehydes (acetaldehyde, propionaldehyde, and butyraldehyde), ketone (acetone, acetoin, diacetyl) and acids (formic acid, acetic acid, propionic acid). Among them, esters and ketones are well known flavourants in wine formed during maturation because of their intense aroma at relatively small concentrations.

4.1.2. **Beer-brewing**

HISTORY: Brewing is the production of malt beverages by yeast fermentation. Beer is believed to be over 10,000 years old. Although no one knows its exact origins, some agricultural historians believe that the first beer may have been produced accidentally when a stash of grain was soaked by rain and then warmed by the sun. If this mixture were spontaneously fermented by wild, airborne yeast—which thrives in just these warm, moist conditions—beer would have been produced. Early beer makers used a very simplistic brewing process and fermented beer for only a brief period, one to two days at most. By around 1100 AD, brewing techniques had become more sophisticated.

Throughout history, wherever cereal grains were grown, humans made a beer-like beverage from them: they used wheat in Mesopotamia, barley in Egypt, millet and sorghum in other parts of Africa, rice in Asia, and corn in the Americas.

Today, beer making is a major industry worldwide. Four basic ingredients are used to brew beer: grain, hops, yeast, and water.

Grain contains the natural sugars required for fermentation. It also provides beer with flavor, color, body, and texture. Hops are small, green, cone-shaped flowers from the hop plant, a vine related to the nettle plant.. Hops provide beer with a spicy, bitter flavour and contribute natural substances that prevent bacteria from spoiling beer. Two species of yeast used to make beer, called brewer's yeast,are *Saccharomyces cerevisiae* and *Saccharomyces uvarum*. Each yeast species is used in a slightly different method of fermentation and produces a distinct type of beer.

Water constitutes as much as 95 percent of the ingredients used in the brewing process. The mineral content in water in particular, the level of salts such as, calcium, sulfate, and chloride dissolved in the water-influences the quality and flavour of the beer it is used to produce.

Steps in beer production are, malting, mashing, separation and boiling of wort, fermentation and maturation.

MALTING: The first step in brewing, called malting, involves steeping the grain in water for several days until it begins to germinate, or sprout Barley grains are allowed to germinate for 4-6 days. During germination, enzymes, α-amylase, β–amylase, proteolytic enzymes within the grain convert the hard, starchy interior of the grain to a type of sugar called maltose. At this point, the grain is called malt. After several days, when the majority of the starch has been converted to sugar, the sprouted grains are heated to stop germination, but without harming the enzymes. This process, called kilning, stops the malt from germinating any further. A portion of the malt may be further roasted to varying depths of colour and flavour to create different styles of beer.

MASHING: The dried malt is processed in a mill, which cracks the husks (the outer coating of the grain). The malt is mixed with malt adjuncts to provide additional carbon substrate. Starchy cereals, maize, rice, wheat (or sugar) are cooked to gelatinize their starch and used as malt adjuncts. The malt and adjuncts are added to water at a ratio of 2:1. The malt adjunct mixture is added with hot water for a period to allow enzymatic conversion at 57-77°C. The malt steeps in the liquid, usually for one to two hours. This process called mashing, breaks down the complex sugars in the grain and releases them in the water, producing sweet liquid called wort. The temperature and amount of time used to mash the malt affects the body and flavour of the finished beer.

SEPARATION AND BOILING OF WORT: The wort is separated from un-dissolved husks and insoluble proteins by filtration. The filtered wort is transferred to a large brew kettle and boiled with agitation for up to two hours. Boiling effectively sterilizes the wort to kill any bacteria that may spoil the wort during fermentation. The hops are added at various times during boiling. Hops are obtained from dried flowers of a plant *Humulus lupulus*. The boiling of wort extracts tannins, essential oils, bitter acids and resins from hops. The tannins help in coagulating un-wanted proteins and thereby protecting the wort from contamination during fermentation and also in establishment of bitter taste and aroma of beer. The essential oils provide aroma and resins are responsible for the pleasant bitterness of beer. At termination of boiling the hops and coagulated materials are removed, the wort is aerated and cooled.

FERMENTATION: The aerated wort is cooled to10°C and placed in a closed fermentation tank- fermentor. The addition of yeast inoculum or pitching of wort is done at 450 g per 120 litres. The anaerobic fermentation of wort is carried out by yeasts, *Saccharomyces cerevisiae* (top yeast), that floats on top of the liquid as it ferments the grain sugars and prefers warm temperatures ranging from 16 to 22°C (60 to 72°F). The fermentation for 40 - 60 hours forms thick surface foam layer. During this time, rapid yeast cell multiplication occurs and evolution of heat increases the temperature to 12-13°C. By fifth day the carbon-dioxide evolution decreases and the foam begins to collapse. From 7 to 9 days the yeast becomes inactive and flocculates and settles to the bottom. The medium is further cooled to hasten settling. The surface scum is removed to help improve flavour of the product. The first fermentation lasts from a few days to two weeks. When the yeast has consumed most of the fermentable sugar, the wort becomes beer. The bottom yeast, *Saccharomyces carlsbergensis,* sinks to the bottom of the liquid and ferments best at cool temperatures ranging from 3 to 10°C (38 to 50°F).

FLOW CHART 4.2. BEER PRODUCTION

MALTING

BARLEY GRAINS SPROUTED: STARCH → MALTOSE

↓

MASHING

↓

DRIED MALT + MALT ADJUNCTS+ HOT WATER → WORT

↓

SEPERATION OF WORT & BOILING

↓

WORT FILTERED+ HOPS - BOILED: EXTRACTS ADDED

↓

FERMENTATION

↓

COOLED WORT +TOP YEAST : 16 -22°C, BOTTOM YEAST 10°C

↓

MATURATION

↓

CONDITIONING TANK-CARBONATED, CHILL POOFING, FILTRATION- CLEAR LIQUID, PASTEURIZATION

Maturation

Maturation lasts for a few weeks to several months, depending on the type of beer being produced. The beer is transferred to an airtight container, called a conditioning tank, for a second fermentation or aging period, where the beer becomes naturally carbonated. Some brewers inject carbon dioxide gas into the beer when aging is complete to give it a bubbly, effervescent quality.

The beer is transferred to storage tanks and held at 0-3°C. Chill proofing is done to remove unstable residual proteins. During this process coagulated components and yeast cells sediment from beer. In addition esters are formed and beer matures losing its harshness. After aging, the beer may appear somewhat cloudy from yeast cells and other particles that remain suspended in the liquid. The most common method of removing these impurities is filtration, a process in which the finished beer is pumped, under pressure, through a sterile filtering system that traps nearly all of the suspended particles from the liquid, resulting in a clear liquid.

Even after filtration, however, some yeast may remain in the beer. To kill the remaining yeast, the beer is pasteurized, or heated to 82°C (180°F) after it has been sealed in cans or bottles. Antioxidants, SO_2 and ascorbic acid are added to prevent oxidative changes in beer which affect flavour. Carbonation and packaging is done before marketing. The ethanol content of beer is 6-8% by volume.

4.1.3. **Bread**

Leavened breads are made by adding baker's yeast to dough made from flour, water and salts. Yeasts ferment sugars in dough to alcohol and carbon dioxide. The carbon dioxide gas bubbles become trapped within the protein matrix of dough and expand in the heat during baking giving the bread its fluffy texture. Dough maturation is accomplished by the chemical reaction of yeast produced alcohols and acids on protein of the flour and by the physical stretching of protein by carbon dioxide gas. This results in light, airy physical structure associated with yeast leavened bakery products. During dough fermentation, yeast produces many secondary metabolites such as ketones, higher alcohols, organic acids, aldehydes and esters. Some of these, like alcohols escape during baking while others react with each other and with other compounds found in dough to form complex flavour compounds. These reactions occur primarily in the crust and the resultant flavour diffuse into the crumb of the baked bread. Flour, water, and leavening agents are the ingredients primarily responsible for the characteristic appearance, texture, and flavour of most bakery products. Eggs, milk, salt, and sugar are effective in modifying these qualities, and various minor ingredients may also be used.

Most of the bakery foods consumed throughout the world are breads and rolls made from yeast-leavened doughs. The yeast-fermentation process leads to the development of desirable flavor and texture, and such products are nutritionally superior to products of the equivalent chemically leavened doughs, since yeast cells themselves add a wide assortment of vitamins and good quality protein. Baker's yeast, *Saccharomyces cerevisiae* is used to leaven bread throughout the world. The scientific name *Saccharomyces,* refers to saccharo meaning sugar and myces meaning fungus. The species name *cerevisiae,* is derived from the name Ceres, the Roman goddess of agriculture. Yeast is essential ingredient in many bakery products and responsible for leavening the dough and imparting a delicious yeast fermentation flavour to the product (Fig. 4.2).

Fig. 4.2 Bread making

Bread Making

Wheat flour (750 g) with a dessert spoonful of salt and sugar is mixed well in a mixing bowl. Ycast cream (inoculum) is prepared by mixing 10 g yeast dried powder with little sugar in water and incubated for 15 minutes. During this time yeast begin to breakdown the sugar and thick creamy foam appear. The yeast cream is poured into flour, stirred gently and water (400 ml) mixed well to make smooth spongy dough. The dough is kneaded by pressing with hands for 10 minutes and leaves the dough for an hour. In this time yeast produces carbon dioxide and the volume of the dough is doubled. The dough is packed in bread tins (10 cm height x 10 cm wide x 20 cm length). The dough in the bread tin is as long and as wide as the tin but about half the height. The bread tins are packed in baking tray, during the time the dough rise to fill the tin. The baking trays are transferred to oven set at 240°C and baked for 30 minutes.

Carbon dioxide is produced by the yeast as a result of the breakdown of fermentable sugars in the dough. The evolution of carbon dioxide causes expansion of the dough as it is trapped within the protein matrix of the dough. The dough maturation is accomplished by the chemical reaction of yeast produced alcohols and acids on protein of the flour and the physical stretching of the protein by carbon dioxide gas. This result in light, airy physical structure associated with yeast leavened products. During dough fermentation, yeast produces many secondary metabolites such as ketones, higher alcohols, organic acids, aldehydes and esters that impart the characteristic flavour. Alcohols escape during baking and others react to form more complex flavour compounds. These reactions occur in the crust and the resultant flavour diffuses in to the crumb of the baked bread.

4.2. Acetic Acid Fermentation: Vinegar

The main desirable fermentation carried out by acetic acid bacteria is the production of vinegar. Vinegar is literally translated as 'sour wine', is one of the oldest products of fermentation used by man. It can be made from almost any fermentable carbohydrate source like fruit, vegetables, syrups and wine. Acetic acid is produced by the oxidation of ethyl alcohol by bacteria such as *Acetobacter aceti, A. orleansis, A. schutgenbachi* and others. The biochemical reactions are as follows:

$$2CH_3CH_2OH + O_2 \rightarrow 2CH_3CHO + 2H_2O$$

$$2CH_3CHO + O_2 \rightarrow 2CH_3COOH$$

Some acetic acid bacteria do not stop at the stage of acid production, but continue the oxidation to carbon dioxide.

$$2CH_3COOH + O_2 \rightarrow 2CO_2 + 2H_2O$$

The production of vinegar depends on a mixed fermentation, which involves both yeasts and bacteria. The fermentation is usually initiated by yeasts which break down glucose to ethanol with the liberation of carbon dioxide. Following on from yeasts, acetobacter oxidize the ethanol to acetic acid and water.

4.3. Lactic Fermentation

Lactic acid bacteria perform an essential role in the preservation and production of wholesome foods. Generally the lactic acid fermentations are low-cost and often little or no heat is required in their preparation. Thus, they are fuel-efficient. Lactic acid fermented foods have an important role in feeding the world's population on every continent today. As world population rises, lactic acid fermentation is expected to become even more important in preserving fresh vegetables, fruits, cereals and legumes for feeding humanity. In milk, the acid coagulates the casein, producing curds. In pickling, the acid preserves the food from pathogenic and putrefactive bacteria.

4.3.1. Sauerkraut

The words Sauerkraut are German terms for sour cabbage. Sauerkraut is a product of natural lactic acid fermentation of fresh cabbage in the presence of a small amount of salt involving lactic acid bacteria (LAB), *Leuconostoc mesenteroides* and *Lactobacillus plantarum*.

Sauerkraut is rich in vitamin C and has a mild laxative action due to the dextran content of the product. A wide variety of sugars, proteins and other nutrients are present in the juices extracted from the cabbage.

Fermentation

The acid forming bacteria are predominant in early fermentation. Soon LAB, *Leuconostoc mesenteroides* predominates in the first stages of pickle fermentation. This species is more resistant to temperature changes and tolerates higher salt concentration than the subsequent species. This organism grows well at 21°C and with 2.5% salt. It out grows all other organisms and continues acid production up to 1% acid. They attack sugars to form lactic acid, acetic acid, ethanol, mannitol, dextran, esters and CO_2, which contribute to the flavour of the product. The final stages of fermentation are effected by *Lactobacillus plantarum* which utilizes mannitol and dextran and removes the bitter taste of Sauerkraut. *Lactobacillus plantarum* continue the production of acid, chiefly produce lactic acid and raise the acidity to 2.5%. The fermentation is stopped by canning or refrigeration of the product.

Production

Fully matured fresh cabbage heads are trimmed, 3 to 4 outer leaves discarded. Cabbage heads are washed, wilted for 2 days,shredded, layered with salt (2.5%), firmly placed in a vat layer by layer, so that brine juice stands on surface and covers to provide anaerobic fermentation condition for 2 to 5 weeks at 25°C. During this period, the acid produced inhibits most microbes except desired LAB. Fermented cabbage is blanched at 70°C for 10 minutes and stored in refrigerator.

4.3.2. **Pickles**

The fermentation of cucumbers is very similar to the sauerkraut fermentation, only brine (salt solution) is used instead of dry salt. The washed cucumbers are placed in large tanks and chlorinated brine with 15 to 20% sodium chloride is added. The cucumbers are submerged in brine ensuring that none float on the surface, to prevent spoilage of pickles. The strong brine drained sugar and water from cucumbers, which simultaneously reduces the salinity of the solution. More salt has to be added to the brine in order to maintain the salt concentration above 12%. This step is essential to prevent the spoilage of the pickles.

Acetic acid and sodium acetate are added to increase the production of acid.

The first the stage of pickle fermentation is predominated by *Leuconostoc mesenteroides* as it is more resistant to temperature changes and tolerates high salinity. As fermentation proceeds and acidity increases lactobacilli start to take over the fermentation. The active stage of fermentation continues for a period of 10 to 30 days depending on the temperature of the fermentation. *Lactobacillus cucumeris* requires an optimum temperature of 29 to 32°C. During fermentation the pH drops to 3.5 and acid content of fermented vegetable increases to 2%.

4.3.3. Traditional Fermented Food-idli

Among the various fermented foods, cereal and legume based fermentation play a vital role in day today diet pattern of India. The traditional methods for preparing cereal and legume based preparations are simple and inexpensive. Rice, pearl millet, finger millet and wheat are the principal cereals and soya bean, mung bean, black gram, Bengal gram are the legumes for the preparation of variety of traditional fermented foods. Traditional foods have evolved through trial and error over many centuries of cultural, social, economic and sensory experience. Conversion of food raw materials into prepared foods of delectable taste is an activity unique to mankind. Traditional food can be described as 'food confined to only a certain geographical locality or ethnicity and which is prepared using raw materials peculiar to the region based on specific process and has been practiced and consumed for generations.' Traditional foods are nutritionally endowed, health promoting and tasty.

Idlies in the form of spongy pancakes are consumed as a breakfast food in south India since, 1100 A.D. Idli is considered a nourishing and easily digestible food. These are prepared by steaming the fermented batter consisting of rice and black gram grains and served with spicy hot side dishes (sambar or chutney). They are auto-fermented foods of acidified and leavened type.

Cereals and legumes are consumed as unprocessed whole grain based preparation in India and have low bioavailability of iron and zinc. Food processing such as soaking, germination and fermentation of whole cereals and legumes activate the endogenous plant and bacterial phytases. These reduce hexa form of phytic acid into lower form, having lower binding ability for metals like iron and zinc. The biochemical changes occur during Indian food fermentations include an increase in volume, total solids, reducing sugars, non-protein nitrogen, soluble nitrogen, amylase, proteinase, inorganic phosphate, free amino acids, including methionine and cystine and water soluble vitamins thiamine, riboflavin, cyanocobalmine. Fermentation also brings about an increase in protein efficiency ratio (PER).

The traditional method of idli preparation consists of washing and soaking parboiled milled rice and black gram dhal in the proportion of 3:1, grinding separately to a fine paste and mixing the two together The mixed paste is allowed to ferment for 12-24 hours at 25-30°C by means of natural microflora originally present in the ingredients.

Both bacteria and yeasts have been reported to participate in idli fermentation and are introduced by the two main ingredients. Black gram dhal soaked in water has a high concentration of soluble nutrients to support the growth of lactic acid bacteria. The bar-boiled rice was found to support the growth of yeasts.

Bacteria identified as part of microflora responsible for the production of good idli include, *Leuconostoc mesenteroides, Lactobacillus fermentum, Lactobacillus delbrueckii, Lactobacillus lactis, Streptococcus faecalis* and *Pediococcus cerevisiae.* The acid and gas production have been found to be mostly dependent on bacterial growth. The role of lactic acid bacteria is to reduce the pH of the batter to a level (pH 4.5) that is optimum for yeast activity. Yeast flora generally involved in the fermentation include, *Saccharomyces cerevisiae, Debaryomyces hansenii, Hansenula anomala, Rhodotorula graminis, Trichosporon beigelii, Torulopsis holmii* and *Torulopsis candida.*

Fermentation brings about two important physical and chemical changes-leavening and acidification in idli batter. These have been used as the criteria for judging the progress of fermentation. The pH decreases from 6.0 to 4.5 and the acidity increases from 3.2 to 19.0 ml of 0.1 N lactic acid per 25 g of idli batter. The volume increases of idli batter are ranging from 1.6 to 3.1 times its original volume. Soluble solids and non-protein nitrogen generally show a slight increase during fermentation, whereas reducing sugars and soluble nitrogen decrease initially and then tend to increase. The microorganisms act on carbohydrate, protein and lipid and produce favourable changes in food. Fermentation increase the amino nitrogen, free sugars, thiamine, riboflavin and inorganic phosphate content as well as sensory qualities of idli.

4.4. MOLD FERMENTATION

Koji Fermented Traditional Foods-sake

Koji mold refers to filamentous molds grown or cultured on solid plant materials in food processing as hydrolytic enzyme producers. It includes Aspergillus molds *Aspergillus oryzae, A.sojae, A. tamari, A. awamori, A. usami* and molds in genera *Rhizopus* and *Mucor.* To make alcoholic beverages and vinegars from starchy raw materials, starch must be converted in to sugars before it is subjected to alcoholic fermentation by yeasts. Traditionally, amylolytic enzymes in germinated grains or malt or koji have been used for the saccharification of starch. Since, ancient times in Asia koji is used in the production of sake (rice wine), Sochu (distilled sake) and rice vinegar.

Sake: The koji, the starter culture for sake brewing, is molded rice grains containing sufficient enzymes, vitamins and various nutritive substances for mashing and for the growth of yeast and fermentation. *Aspergillus oryzae* is cultured on steamed rice at 36°C for 5 days. In the first step steamed rice, koji and water are added to seed mash. After two days, the yeast population increases and the second addition of material is made. Next day third addition is done. After 15 to 20 days, the mash is squeezed, the filtrate is clarified and pasteurized. It is stored for about 3 to 8 months and then filtered, diluted to 15% alcohol content, bottled and pasteurized.

Enzymatic degradation of proteinaceous foods into digestible foods or tasty condiments in East and South East Asian countries is generally done with koji molds. Douchi is prepared from molded beans using *Aspergillus oryzae.*

TEMPEH: Tempeh is a white mold covered, compact cake produced by fermentation of dehulled, hydrated, soaked and pre-cooked soya bean cotyledons. Tempeh occupies a prominent place among popular fermented foods in South Asian region. Soya beans are soaked, dehulled and partially cooked. This is followed by surface drying, inoculation with *Rhizopus oligosporus* and *Rhizopus oryzae* spores and allowing to ferment semi-aerobically for about 36 hours to get a compact moldy white solid product. The mycelium of the mold knits the cotyledons into a compact cake that can be sliced and cut into cubes and used as protein rich meat substitute. The process of tempeh preparation makes soyabean more digestible and lowering cholesterol and phytic content. Beside, tempeh contains vitamin B_{12}, enhanced isoflavanoids as anti-oxidants and more available iron.

SOY SAUCE: Savory flavoured sauces are also traditionally prepared by fermentation of soyabeans, which transform bland vegetable protein into meat flavoured sauces. In the traditional method soyabeans are soaked and thoroughly cooked, mashed and formed into balls. These are covered with straw and hung to ferment under the rafters. In 30 days, they completely overgrow with the mold *Aspergillus oryzae.* The fermented mass is placed in a strong salt brine (18% w/v) allowing the beans to be hydrolyzed by the proteases, lipases and amylases in the mold. As a result of this enzymatic action, a liquid sauce with a meaty flavour is obtained.

MISO is a semisolid salty food prepared using koji mold. Rice miso is produced using rice and cooked soya beans, Barley miso from barley and cooked soya beans and soya bean miso from cooked soya beans.

Traditional fermentations are likely to remain an important part of global food supply as it has advantages like utilization of local raw material, inexpensive and familiar process method. Additional research work on genetic improvement of microbes, optimization of favourable conditions for quicker fermentation, improvement in quality and quantity of fermented foods and safety in use are required for commercialization of fermented foods.

QUESTIONS

1. Define femented food.
2. What are the types of femented food and the microbes involved?
3. What are the types of wine?
4. Describe the wine making with the help of a flow diagram.
5. Write a brief account on aging of wine.
6. What is Brewing? Describe the steps involved in beer production.
7. Explain the role of hops in brewing.
8. Give an account on bread making.
9. What is Sauerkraut? Explain the Sauerkraut fermentation.
10. Describe the nature of idli fermentation with the organisms involved.
11. Short notes on : wort, must, top yeast, bottom yeast, koji mold.
12. Malt is prepared from

    a) Wheat b) Barley

    c) Corn d) Rice

13. In beer production the liquid to which yeasts are added is called
    - a) Must
    - b) Wort
    - c) Malt
    - d) Lager
14. Which of the following foods is not a product of microbial fermentation?
    - a) Brandy
    - b) Cheese
    - c) Soy sauce
    - d) All of the above
15. Sauerkraut is produced under what type of conditions?
    - a) Aerobic
    - b) Anaerobic
    - c) Microaerophilic
    - d) It doesn't matter
16. A yeast's ability to produce naturally leavened bread depends largely on what substance in wheat flour?
    - a) Casein
    - b) Lactose
    - c) Gluten
    - d) Wort

5

DAIRY MICROBIOLOGY

Milk is nutritious and essential food for human beings and also serve as good medium for microbial growth and contamination. Milk is a single, most nearly complete, natural food. Primarily it is meant for nourishing the newborns but man has exploited it for human consumption since long. Today milk has become indispensable integral component of our diet. About half the milk produced is consumed as fresh milk, the rest going into a variety of products including evaporated and condensed milk, dried milk powder, butter, cheese, malted milk, ice cream, yoghurt, lactose and casein. Milk is very important from a nutritional standpoint since it contained all the essential food constituents (carbohydrate-lactose, protein-casein, fat-butter, minerals-calcium, potassium and vitamins- A, B, C, D, E, and K) required in human diet in about the proper proportions. India is the second largest producer of milk in the world. India has exhibited unprecedented rate of growth in dairy development in last 30 years mainly due to sustained efforts of dairy technologists in Operation flood Programme introduced by the National Dairy Development Board. Dairy microbiology deals with aspects like shelf life and spoilage microbes, public health significance and food borne pathogens, and production of fermented dairy products.

5.1. MICROORGANISMS IN MILK

SPOILAGE MICROBES: Spoilage is a term used to describe the deterioration of foods texture, colour, odour or flavour to the point where it is unsuitable for human consumption. Microbial spoilage of food often involves the degradation of proteins, carbohydrates and fats by the microbes or their enzymes. The microbial quality of milk is crucial for the production of quality dairy food products. In milk, the microbes that are principally involved in spoilage are psychrotrophic organisms. Most psychrotrophs are destroyed by pasteurization temperatures. However, *Pseudomonas fluorescens, Pseudomonas fragi* can produce proteolytic and lipolytic extra-cellular enzymes which are heat stable and capable of causing spoilage. Few species of *Bacillus, Clostridium, Corynebacterium, Arthrobacter, Lactobacillus, Microbacterium, Micrococcus* and *Streptococcus* can survive pasteurization and can grow at refrigeration temperatures which can cause spoilage problems. The common bacteria found in milk are: rickettsia- *Coxiella burnetti,* lactic acid bacteria- *Lactobacillus bulgaris, Streptococcus thermophilus,* aerobic rods- *Bacillus subtilis, Bacillus cereus* and anaerobic bacteria- *Clostridium butyricum.*

COLIFORMS: Coliform are facultative anaerobes with an optimum growth at 37°C. Coliforms are indicator organisms, they are closely associated with the presence of pathogenic organisms but not necessarily pathogenic themselves. They also can cause spoilage of milk because they are able to ferment lactose with the production of acid and gas and able to degrade milk proteins. They are killed by HTST treatment.

PATHOGENIC MICROBES: Certain diseases may be transmitted through milk, from a cow. It is essential that cows producing milk should be free from any contagious or infectious disease. Bovine tuberculosis comes from consumption of milk of tuberculous cows. Tuberculosis, brucellosis, mastitis and actinomycosis are the human disease transmitted through milk. The disease producing bacteria may find their way into milk and cause infectious diseases in children and adults.

Hygienic milk production, proper handling and storage of milk, and mandatory pasteurization have decreased the threat of milk-borne diseases such as tuberculosis (*Mycobacterium tuberculosis*), brucellosis (*Brucella melentensis*) and typhoid fever (*Salmonella typhi*). There have been a number of food-borne diseases resulting from the ingestion of raw milk or dairy products made with milk that was not properly pasteurized or poorly handled causing post processing contamination. The following bacterial diseases are transmitted through raw milk and other dairy products: Q Fever- *Coxiella burnetii,* Diphtheria- *Corynebacterium diphtheriae,* Typhoid- *Salmonella typhi,* Tuberculosis- *M.tuberculosis,* Bovine TB- *M.tuberculosis* var *bovis,* Gastrointestinal disorders-*Shigella, Salmonella, Bacillus cereus, Listeria monocytogenes, Yersinia enterocolitica, E-coli* 0157-H7, *Campylobacter jejuni.*

Molds, mainly species of *Aspergillus, Penicillium,* and *Fusarium* can grow in milk and dairy products. Under suitable conditions the molds may produce mycotoxins, which can be a health hazard in dairy products. Contaminated milk reduces the chances of high quality production of dairy products and thereby affects the human health and economy badly.

5.2. FERMENTED DAIRY PRODUCTS

Lactic acid bacteria have been used for at least for 4000 years, to ferment foods, particularly fermented milk products from all over the world, including cheese, yoghurt, butter, butter milk, kefir and koumiss. Although they are best known for their role in the preparation of fermented dairy products, they are also used for pickling of vegetables, baking, wine making, curing fish, meats and sausages. They are a large group of beneficial bacteria wide spread in nature and also found in human digestive system. Lactic acid bacteria produce lactic acid as end product of fermentation process. The milk sugar lactose is converted to lactic acid and as the acid accumulates, the structure of the milk protein changes (curdling). Lactic acid also gives fermented milks their slightly tart taste. Additional characteristic flavours and aromas are often the result of other products of lactic acid bacteria. There are considerable opportunities for exploiting lactic acid bacteria as probiotic cultures. These supplement and help our normal gut flora to function more efficiently. Lactic acid bacteria are therefore excellent ambassadors for an often maligned microbial world. They are not only of major economic significance, but are also of value in maintaining and promoting human health.

LACTIC ACID BACTERIA: Lactic acid bacteria (Fig. 5.1) ferment sugars to lactic acid . They are normally present in milk, and are also used as starter cultures in the production of cultured dairy products. Their activity is desirable in a variety of foods such as dairy products and sauerkraut, for the production of flavour. Common homolactic bacteria include *Lactobacillus delbrueckii, Lactobacillus plantarum, Streptococcus faecalis, Pediococcus cerevisiae, Lactobacillus lactis* and heterolactic bacteria include *Leuconostoc mesenteroides Lactobacillus fermentum. Lactobacillus cremoris, Lactobacillus brevis.* These bacteria produce organic acids, diacetyl (aroma compounds), bacteriocins, exo-polysaccharides, hydrogen peroxide and several enzymes during fermentation.

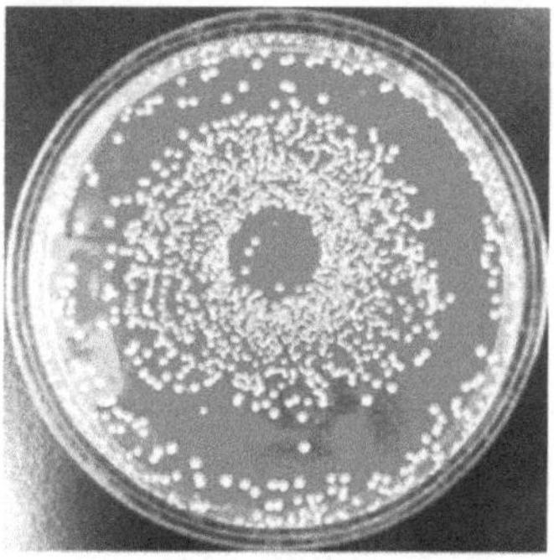

Fig. 5.1. Colonies lactic acid bacteria

STARTER CULTURES: Starter cultures are those microorganisms that are used in the production of cultured dairy products such as cheese and yoghurt. The natural microflora of milk is either, inefficient, uncontrollable and unpredictable or is destroyed during pasteurization.

A starter culture can provide particular characteristics in more controlled and predictable fermentation. The primary function of lactic starters is the production of lactic acid from lactose. The other functions are flavour, aroma and alcohol production; proteolytic and lipolytic activities; inhibition of undesirable organisms.

STARTER CULTURES USED FOR FERMENTED DAIRY PRODUCTS

PRODUCT	CULTURES RESPONSIBLE FOR FERMENTATION
YOGHURT	*Streptococcus thermophilus, Lactobacillus delbrueckii* subsp *bulgaricus*
DAHI	Lactococci, *Leuconostoc, Lactobacillus delbrueckii* subsp.*bulgaricus Lactobacillus acidophilus, S. thermophilus*
ACIDOPHILUS MILK	*Lactobacillus acidophilus*
BULGARIAN MILK	*Lactobacillus delbrueckii* subsp. *bulgaricus*
CULTURED BUTTER MILK	*Streptococcus lactis, S. cremoris, Leuconostoc cremoris Leuc. dextranicum*
KEFIR	*KEFIR GRAINS* Lactobacilli, yeasts, Lactococci, Acetic acid bacteria
KOUMISS	*Lactob. delbrueckii* subsp. *Bulgaricus, Lactob. acidophilus, Kluyveromyces fragilis*
SWISS CHEESE	Thermophilic lactobacilli-*Streptococcus thermophilus, Propionibacterium freudenreichi* subsp. *shermanii*
CHEDDAR CHEESE	Lactococci, *Leuconostoc mesenteroides* subsp. *cremoris*
BRICK CHEESE	Lactococci, Brevibacterium *linens*
ROQUEFORT CHEESE	Lactococci, *Penicillium roqueforti*
CAMEMBERT CHEESE	Lactococci, *Penicillium camemberti*

MESOPHILIC STARTER CULTURES

Lactococcus lactis subsp *cremoris*

Lactococcus lactis subsp *lactis* biovar *diacetylactis*

Lactobacillus delbrueckii subsp *lactis*

Leuconostoc mesenteroides subsp *cremoris*

THERMOPHILIC STARTER CULTURES

Streptococcus thermophilus

Lactobacillus delbrueckii subsp *bulgaricus*

Lactobacillus caesi

Lactobacillus plantarum

Lactobacillus helveticus

5.3. FERMENTED DAIRY PRODUCTS

1. **Yoghurt:** Yoghurt, a fermented, semifluid milk product, is used extensively as a health food. It is prepared from fresh whole or skim milk, boiled and concentrated by evaporation. Fermentation is caused by the addition of mixed culture of lactic acid bacteria, usually *Streptococcus thermophilus* and *Lactobacillus bulgaricus* (sometimes *Lactobacillus acidophilus*) in a ratio of 1:1.

   The fermentation is carriedout with in a temperature of 35 to 45°C. The coccus grows faster than rod and is primarily responsible for acid production, while the rod adds flavour and aroma. The associative growth of the two organisms results in lactic acid production at a rate greater than that produced by either when growing alone. The chief volatile flavour component of yoghurt, acetaldehyde is produced by *Lactobacillus bulgaricus,* when growing in association with *Streptococcus thermophilus.* New batches of yoghurt (Fig. 5.2) can be produced by introducing into concentrated milk a portion of a previously prepared batch. This type of fermented milk has long been an important constituent of the diet in south Eastern Europe and in Asia Minor.

Fig. 5.2. Lactic acid bacteria *Streptococcus thermophilus* L. bulgaricus in Yoghurt

2. **Buttermilk** is not high in fat as the name suggest. Actually, the name refers to the fact that buttermilk was once the watery end-product of butter making. Modern buttermilk is made from low-fat or skim milk and has less than 2 percent fat and sometimes none. Its correct name in many jurisdictions is "cultured low-fat milk or "cultured nonfat milk." The starting ingredient for buttermilk is

skim or low-fat milk. The milk is pasteurized at 82 to 88°C for 30 minutes, or at 90°C for two to three minutes. This heating process is done to destroy all naturally occurring bacteria and to denature the protein in order to minimize wheying off (separation of liquid from solids). The milk is then cooled to 22°C, and starter cultures of desirable bacteria, such as *Streptococcus lactis, S. cremoris, Leuconostoc citrovorum,* and *L. dextranicum,* are added to develop buttermilk's acidity and unique flavour. These organisms may be used singly or in combination to obtain the desired flavour. The ripening process takes about 12 to 14 hours (overnight). At the correct stage of acid and flavour, the product is gently stirred to break the curd, and it is cooled to 7.2°C in order to halt fermentation. It is then packaged and refrigerated.

3. **Cultured butter milk:** Cultured butter milk is fresh skimmed pasteurized milk that has been fermented by lactic culture and aroma organisms. The product obtained is a drinkable viscous fluid containing characteristic pleasing aroma and flavour. The bacteria used in fermentation are selected strains of *Streptococcus cremoris* in combination with *Leuconostoc mesenteroides* subsp *cremoris.* The skim milk is heated at 82-87°C for 30 minutes, cooled to 24°C and inoculated with starter. The heat treatment of milk destroys undesirable pathogenic and spoilage microflora, eliminates any biological competition for starter bacteria, destroys natural inhibitors in milk and facilitates partial denaturation of whey proteins which makes them susceptible to precipitation and coagulation as the pH value drops during fermentation.

4. **Sour cream:** Sour cream results from souring by lactic acid producing bacteria of pasteurized cream. Sour cream contains not less than 18% milk fat. Sour cream is used as lopping on baked potatoes, vegetables, salads and fruits. Sour cream is a companion product for cultured butter milk because the starter bacteria and culturing conditions are similar.

5. **Ripened cream butter:** Ripened cream butter is the product obtained by churning cream that has been cultured with selected strains of lactic streptococci and leuconostocs. Ripened cream butter has a much better flavour than sweet cream butter because of the pleasant lactic acid flavour and diacetyl derived from the ripened cream.

6. **Cheese:** Cheese is a fermented dairy product made from curd, obtained by the coagulation of milk casein with an enzyme rennin or lactic acid. The curd is further processed with heat, pressure, salt and ripening with selected microorganism to obtain cheese. Cheese,

nutritious food made from the milk of cows and other mammals, including sheep, goats, buffalo, reindeer, camels, yaks, and mares. Cheese is one of the world's oldest food products-for thousands of years, people have raised animals for milk, turning their surplus milk into cheese. More than 400 varieties of cheese exist, making it one of the most versatile foods in the world. Cheese comes in hundreds of different shapes, sizes, textures, flavours, and aromas and is used in as many different ways. Enjoyed with bread, crackers, and fruit, used as an ingredient in cooked foods, or grated on salads and pastas, cheese is a healthy component of cuisines all over the world.

Cheese (Fig. 5.3) is a concentrated source of almost all the valuable nutrients found in milk, such as protein, vitamins, and minerals, as well as the less of undesirable fat and cholesterol, substances that may lead to health problems when consumed in excess. The fat content in cheese varies depending on the milk used. Cheese made with whole milk, or milk enriched with cream, has the highest amount of fat, cholesterol, and calories. Cheese made with skim milk has the lowest. Because of its high protein and calcium content, cheese is an important component of a balanced diet. It is an especially good source of protein for children, whose growing bodies require higher amounts of protein than adults. Many vegetarians, who do not eat meat, rely on cheese as a source of protein in their diets.

Fig. 5.3 Cheese

There are three **basic steps** common to all cheese making. First, proteins in milk are transformed into solid lumps called curds. Second, the curds are separated from the milky liquid, called whey, and shaped or pressed into moulds. Finally, the shaped curds are ripened using a variety of different aging and curing techniques.

Curd making: Curds are formed when an enzyme called rennin is stirred into milk. Rennin encourages casein, one of the proteins in milk, to solidify and clump together. Rennin is found in rennet, a substance extracted from the stomach linings of calves, sheep, and goats, and also in certain plant extracts. Rennin aids coagulation only if the milk is slightly acidic, as it is when it becomes sour. The cheese makers speedup the process by warming the milk and adding specialized bacteria that convert the sugars found in milk to lactic acid, creating the acidic environment necessary for casein coagulation. As the casein clumps together, it traps fat globules and some of the milky liquid inside the clumps, forming moist, nutritious curds.

Separation of whey: In the second step of cheese making, the curds are separated from the whey. The curds are cut into small chunks to drain the excess whey trapped inside. Different types of cheeses have varying moisture contents, determined by the amounts of whey allowed to remain in the curds. To produce cheeses high in moisture content, such as cottage cheese, cheese makers need only cut the curds and drain the whey before seasoning and packing the cheese into cartons for sale. Cheeses with lower moisture content undergo further treatment to condense the curds and remove more of the whey. The varying methods used to release excess whey play a large role in determining the final character of the cheese. Some cheeses, such as Swiss cheese, are heated and agitated in a cooking process that further breaks down the curds and releases more of the whey, creating a denser style of cheese. To create cheeses that are denser still, the curds are stacked on top of each other, in a process called cheddaring, which relies on the weight of the layers to squeeze still more of the whey from the curd. At this point, cheese makers may knead the curds-twisting and pulling them by hand-to create a stringier, more elastic texture.

The curds are then shaped by hand or pressed into moulds of various shapes and sizes. Curds of nearly all cheeses are salted by stirring the salt directly into the curds or by rubbing salt or a saltwater solution, called brine, onto the curd surface. Salt pulls moisture from the cheese, but more importantly, it acts as a preservative and slows down the final step of cheese making-the ripening.

Ripening: During the ripening process, microbes such as bacteria slowly change the composition of the curds, creating cheeses with distinct flavours, textures, and aromas. The kinds of microbes used, the temperature and humidity conditions of the ripening environment, and the duration of the ripening process, all contribute to the final characteristics of the cheese. In some cheeses, the bacteria added to create the acidic environment necessary for curd formation continue to ripen the cheese as well. In Swiss cheese, for example, these bacteria produce gas bubbles during ripening, creating its characteristic holes, or eyes. In other cases, microbes are added to the shaped curd. For example, a blue-green mold called *Penicillium roqueforti* is used to ripen cheeses such as Roquefort and Gorgonzola. This special mold creates bluish-green veins in the cheese and a characteristic sharp flavour and creamy texture. Other cheeses, such as Brie and Camembert, are ripened by bacteria rubbed on the outer surface of the cheese. The bacteria slowly work their way into the interior of the cheese, creating a soft, pungent interior and leaving a powdery, edible white rind on the outside.

Temperatures are kept cool, not only to encourage the activity of the ripening bacteria but to inhibit the growth of harmful bacteria that could spoil the cheese. The amount of time that cheeses are allowed to ripen, or age, also contributes to their final character. Generally, cheeses develop sharper flavours, harder textures, and deeper colours as they age.

CHEESE TYPES: Based on the moisture content, they are classified in to three types (Table 5.1) and Based on the ripening, they are classified in to three types (Table 5.2).

TABLE 5.1 CHEESE TYPES BASED ON MOISTURE CONTENT

TYPE	MOISTURE CONTENT	EXAMPLE
Hard cheese	< 35 % moisture	Cheddar cheese
Semi hard cheese	35-50 % moisture	Blue cheese, roquefort cheese
Soft cheese	>50% moisture	Camembert, Cottage cheese

TABLE 5.2 CHEESE TYPES BASED ON RIPENING

TYPE	EXAMPLE	ORGANISM
Unripened cheese	Cottage cheese	Soft white cheese made from curds of skim milk and ripened cheese
Ripened cheese with bacteria	Swiss cheese	*Propionibacterium shermanii*
Ripened cheese with fungi	Camembert cheese Blue cheese	*Penicillium camemberti* *Penicillium roqueforti*

Cottage cheese: Cottage cheese is a soft cheese, generally coagulated with lactic acid rather than rennin. The curd is not pressed, aged or ripened. Starting with pasteurized skim milk, curd forming operations have many similarities to the early stages of all cheese making.

Swiss cheese is a hard type cheese, characterized by large holes or eyes and sweet nutty flavour, obtained through the activities of *Propionibacterium shermani*. This organism follows the growth of LAB and ferments lactic acid to propionic acid. The propionic acid contributes to the nutty flavour and carbon dioxide gas collects in pockets within the ripening curd and forms the eyes.

STAGES IN CHEDDAR CHEESE PRODUCTION: **Curd making:** Pasteurized milk with 3.5% milk fat is adjusted to 32°Cand inoculated with 2% lactic starter culture *Streptococcus lactis* and allowed for 30 minutes ripening. Then rennet extract is added and stirred. **Cutting of curd:** Curd is cut into cubes of ½″ (1.5 cm) sides to ensure efficient removal of whey. **Cooking:** Curd cubes are gently agitated and heated in vat using steam to 38°C for 30 minutes. Heat increases the rate of acid formation and makes the cubes to shrink. **Matting and cheddaring:** Matting involves fusion of curd pieces to form rubbery slab. Cheddaring involves cutting of matted curd into blocks and piling the blocks on one another to squeeze out whey from the curd. **Milling and salting:** Curd blocks are milled into pieces of 2" long and spread out on a vat and sprinkled with 2.5% salt (2.5 kg/100 kg of curd). Salt has three functions - to draw out whey away from the curd by osmosis, to act as preservative and to add flavour. **Pressing:** Salted curd pieces are pressed in a hydraulic press at 20 psi over night. Pressing determines the final shape and texture of the product. **Curing and ripening:** After pressing, the cheese is placed in a cool drying room at 15°C and at 60% RH for three days. This causes mild surface drying and formation of a slight rind. Cheese is wax coated to prevent mold growth and excessive drying during curing period of 60 days at 2°C and 85% RH. During curing period the bacteria in cheese and enzymes in rennet modify cheese texture, flavour and colour by continuing to ferment residual lactose to acids and other aroma components.

Cheese-making technology developed rapidly during the last half of the 19th century. Food safety was increased when French microbiologist Louis Pasteur proved that harmful organisms in raw milk are killed when it is heated to 62°C for 30 minutes, or to 71°C for 15 seconds, in a process called pasteurization. Many cheeses are now made from pasteurized milk, decreasing the likelihood that harmful bacteria will contaminate cheese during the ripening process.

Today, most of the world's cheese is mass-produced in large commercial factories that employ industrial equipment and state-of-the-art technology. The European Union ranked first in annual cheese production and consumption in 2003, and the United States ranked second.

FERMENTED MILKS: Kefir and Koumiss

These are acid-alcoholic fermented milks with high therapeutic value, popular in Russia and Central Asia. Kefir is made from whole milk of goats, sheep or cows held at 85°C for 30 minutes. The heated milk is cooled to 22°C, inoculated with kefir grains and incubated over night at 22°C. The resultant smooth curd is strained through a wire sieve to recover the kefir grains. The kefir grains are whitish or yellowish gelatinous irregular granules of varying size and largely made up of a polysaccharide called kefiran. The granules are insoluble in water and swell up when soaked in water to form a jelly. Within the folds of granules the mixture of at least six functionally different groups of bacteria (*Lactobacillus, Streptococcus, Micrococcus, Bacillus*) and yeasts (*Saccharomyces kefir, Torula kefir, Candida*) are found. Kefir grains recovered from a batch of kefir can be dried in room temperature for 36 hours and they retain activity for 12-18 months. The product has 1.0 per cent lactic acid, 0.5 per cent alcohol and 0.05 per cent carbon dioxide.

Koumiss is traditionally made from fresh mares milk warmed to 28°C. The milk is then inoculated with 30% of a starter consisting lactic acid bacteria (*Lactobacillus delbrueckii* subsp *bulgaricus*) and yeasts (*Torula kefir*). The culture is prepared by growing each organism separately in cow's milk, then adding both to mare's milk. The inoculated mare's milk is agitated to facilitate the growth of yeast. A product with sufficient acidity is obtained in 2 hours incubation. Koumiss is bottled warm, capped and incubated for an additional 2 hours at 20°C. It is cooled to 4°C and held refrigerated until consumed. The finished koumiss is naturally carbonated and contains 0.8 -1.6 % lactic acid and 0.7-2.5 % ethanol. The product is graded as weak, medium and strong based on ripening period and acid-alcohol content.

5.3. PROBIOTICS

The term "probiotic," which literally means "for life," has since been employed to describe these health-promoting bacteria. The World Health Organization has defined probiotic bacteria as "live microorganisms which when administrated in adequate amounts confer a health benefit on the host by improving the intestinal microbial balance". Probiotic bacteria have been historically used to treat a variety of ailments, including infections of mucosal surfaces such as the vagina and the gastrointestinal tract.

Bacteria employed in probiotic applications help to maintain or restore a host's natural microbial flora. The ability of probiotic bacteria to successfully outcompete undesired species is often due to, or enhanced by, the production of potent antimicrobial toxins. The most commonly encountered of these are bacteriocins, a large and functionally diverse family of antimicrobials found in all major lineages of bacteria. Recent studies reveal that these proteinaceous toxins play a critical role in mediating competitive dynamics between bacterial strains and closely related species. The potential use of bacteriocin-producing strains as probiotic and bio-protective agents has recently received increased attention.

In 1908, Elie Metchnikoff, working at the Pasteur Institute, observed that a surprising number of people in Bulgaria lived more than 100 years. This longevity could not be attributed to the impact of modern medicine because Bulgaria, one of the poorest countries in Europe at the time, had not yet benefited from such life extending medical advances. He further observed that Bulgarian peasants consumed large quantities of yogurt. He subsequently isolated bacteria from the yogurt and determined that they conferred the observed health-promoting benefits. Nearly a century elapsed before mainstream health providers considered using such bacteria to improve the health of their patients.

Two main genera of lactic acid bacteria dominate the intestinal flora, including 56 species of *Lactobacillus* and numerous species of *Bifidobacterium*. Most of these species have been shown to produce bacteriocins *in vitro* and *in vivo*.

Desirable properties of Probiotics: Consumers are attracted towards probiotics as they impart beneficial properties without any genetic or molecular changes.

1. Resistant to acid and bile toxicity.
2. Adherence to human intestine cells.
3. Colonization in human guts.
4. Antagonism against pathogenic bacteria.
5. Production of antimicrobial substances.
6. Immune modulation properties.

Mechanism of Probiotics: Unfavourable environment for pathogens is created by these mechanisms.

1. Production of antimicrobial compounds: bacteriocins, short chain fatty acids, organic acids, hydrogen peroxide, dicetyl, reuterin,
2. Competition for nutrients.
3. Reduction of pH.
4. Immune tolerance.
5. Blockage of adherence of entero-pathogenic bacteria, viruses, parasites, by occupying the niche of intestinal mucosa.

LANTIBIOTICS: Bacteriocins produced by LAB, which have a long history of use in fermentation and meat and milk preservation, are the best characterized of this group. Lactic acid bacteria have been employed for centuries in the fermentation of food, partly due to the fact that they can prevent the growth of spoilage and pathogenic microorganisms. They produce bacteriocins called lantibiotics-Lantibiotics are ribosomally synthesized bacteriocins that target a broad range of Gram-positive bacteria. Type A lantibiotics, such as nisin, are small (2–5 kDa), elongated, screw- shaped proteins that contain positively charged molecules, which kill via the formation of pores, leading to the dissipation of membrane potential and the efflux of small metabolites from the sensitive cells. Nisins have a dual mode of action: (1) They bind to lipid II, the main transporter of peptidoglycan subunits from the cytoplasm to the cell wall, and therefore prevent correct cell wall synthesis, leading to cell death, and (2) they employ lipid II as a docking molecule to initiate a process of membrane insertion and pore formation that leads to rapid cell death.

Nisin A is shown to kill a wide range of organisms including *Actinomyces, Bacillus, Clostridium, Corynebacterium, Enterococcus, Gardnerella, Listeria, Micrococcus, Mycobacterium, Propionibacterium, Streptococcus and Staphylococcus.* Contrary to conventional wisdom, these bacteriocins are also active against a number of medically important Gram-negative bacteria including *Campylobacter, Haemophilus, Helicobacter,* and *Neisseria.* One weakness of the bacteriocins produced by Gram-positive bacteria, with respect to their use in probiotic applications, is that they seldom inhibit commonly encountered enteropathogenic bacteria such as *Enterobacter, Klebsiella, or Salmonella.* However, bacteriocins produced by Gram-negative bacteria can accomplish this task. Microcins produced by *E. coli* hold promise in reducing the abundance of *Salmonella typhimurium.*

APPLICATIONS: Probiotic organisms, the lactic acid bacteria, favour and benefit the biological systems of the other organisms by many ways.

1. They produce preservative compounds such as organic acids-lactic acid, acetaldehyde, hydrogen peroxide, diacetyl and bacteriocins. Nisin is the best known bacteriocin widely used in various foods such as canned foods, dairy products, meat products. Bacteriocins are ribosomally synthesized, extracellularly released bioactive peptides which have a bactericidal or bacteriostatic effect on other closely related species. Nisin, lactin, lactocin are bacteriocins produced by lactic acid bacteria and used in milk and sausages

2. Suppress the food pathogens, allergens, toxins and spoilage organisms.
3. Beneficially affect host animal by improving its microbial balance-used as microbial dietary adjuvant that beneficially affect host physiology.
4. Improve nutrition and acceptability of fermented foods. No undesirable side effects
5. Enhance shelf life of fermented foods with supportive preservation by reduction of initial microbial count-hurdle effect.
6. Cost effective technology and energy saving method of preservation-through fermentation.
7. Possess therapeutic uses for many gastrointestinal diseases.

The cultures of *L. plantarum* and *Bifidobacteruim adolescentis* exhibited a broad spectrum antibacterial activity against selected food borne pathogenic bacterial species.

QUESTIONS

1. What are the types of microorganisms found in milk?
2. What are the pathogenic organisms transmitted through milk?
3. What are the various types of cheese?
4. Describe the basic steps in cheese making
5. What is starter culture? List out various starter cultures of fermented dairy products.
6. What are the characteristics of probiotics?
7. What are lantibiotics?
8. Write short notes on : yoghurt, Kefir and Koumiss
9. The holes in swiss cheese are a result of microbial production of

   a) O_2 b) CO_2

   c) Lactic acid d) Acetic acid
10. An example for hard ripened cheese is

    a) Cottage cheese b) Cheddar cheese

    c) Camembert cheese d) Blue cheese

❑❑❑

6

MICROBIAL FOODS (SCP) AND FOOD CHEMICALS

SINGLE CELL PROTEIN (SCP) AND PROTEIN DEFICIENCY

Protein is an essential component of diet. The greatest single problem in the world today is global food protein shortage. Having adequate quantity of food is only one side of the picture, but the most important aspect is the supply of balanced and nutritious food. All efforts are being made to increase the food production through modern methods of farming. But very little attention is being paid towards protein requirements and production. Approximately half of world population suffers from the protein deficiency disease called 'Kwashiorkor' and children with poor protein are victims of this disease.

Usually, cereals have low protein values while protein rich pulses are not available in desired quantities due to poor yields and high costs. The good quality protein available in fish, meat, poultry, and eggs is not available due to poor economic condition as well as varied dietary habits. With the current system of production, agriculture cannot be relied upon to feed an ever increasing world population. Hence there is an urgent need to find other protein sources. The best potential is seen in microbial protein or single cell protein (SCP), a new source of protein independent of agriculture.

DEFINITION: The dried cells of microorganisms such as bacteria, fungi, yeasts, algae that are grown in large scale culture systems as proteins for human or animal consumption are collectively known as "single cell proteins".

SCP is the protein extracted from cultivated microbial biomass. It can be used for protein supplementation of a staple diet by replacing costly conventional sources to alleviate the problem of protein scarcity. Moreover, bioconversion of agricultural and industrial wastes to protein-rich food and fodder stocks has an additional benefit of making the final product cheaper. This would also offset the negative cost value of wastes used as substrate to yield SCP. Further, it would make food production less dependent upon land and relieve the pressure on agriculture.

The recent advances in microbiology and biotechnology have made the approach of microbial protein production on industrial scale feasible. The protein advisory group of UNO has approved SCP as an animal feed supplement and as a human dietary constituent. In animal food it is replacement for traditional supplements such as soybean meal and fish meal. For humans, the SCP is used either as a food additive to improve product quality or as a protein supplement.

The main advantages of SCP-over plant and animal proteins are:

1. Microbes synthesize proteins much more rapidly than plants and animals. They are capable of doubling their biomass in few hours, whereas plants take few weeks and animals few months.
2. The protein content of microbes is higher, 43-85% in dry mass, than that in common foods.
3. SCP can be produced in a continuous culture, independent of climate and with low land and water requirements.
4. Microbial proteins can be produced on raw materials locally available which also include waste products. Thereby proteins are produced from wastes and the waste recycling prevents pollution in the environment.
5. The microbes can be genetically modified for desirable characters such as amino acid composition, rapid growth rate, heat resistance.

6.1. MICROBIAL GROUPS AND SUBSTRATES

The range of suitable taxonomic groups of microorganisms and of substrates for these processes has lately been broadened (Table 6.1). Yeast is probably the most widely accepted and used microorganisms for Single Cell Protein production. These include strains of *Candida utilis, Candida arborea, Candida pulcherrima* and *Saccharomyces cerevisiae.* Attention has been focused on the use of renewable feedstock and of industrial wastes (cellulose-containing materials, etc.). Use of the latter for microbiologic synthesis could produce enough protein to make up for the global deficit in protein. At present, production of SCP by mass culture is in infancy. It needs much boost to solve the problem of starvation in the new millennium. One of the ways to improve quality and enhance productivity of SCP product is the genetic improvement of the microbes. Transfer and expression of beneficial genes through genetic engineering in microbes have opened a new era for the mass production of SCP.

Table 6.1 TYPES OF MICROORGANISMS EMPLOYED FOR SCP PRODUCION

GROUP	MICROORGANISM	PROTEIN %	SUBSTRATE
Bacteria	*Methylophilus methlyotrophus*	72	Methanol
	Methylomonas sp	70	Methanol
	Pseudomonas sp.5401	62	n-paraffins,fuel oil
	Cellulomonas sp.	87	Bagasse
Yeasts	*Torulopsis sp.*	50	Methanol
	Candida utilis		54 Molasses, milk whey
	Saccharomyces cerevisiae	53	Molasses
Molds	*Fusarim graminarium*	50	Hydrolyzed starch
	Trichoderma viride	54	Straw, starch
	Paecilomyces varioti		55 Sulphite waste liquor
	Penicillium cyclopium	55	Cheese whey
Algae	*Chlorella pyrenoidosa*	50	CO_2, sunlight
	Spirulina maxima	60	Mineral medium, sunlight
	Spirulina platensis	63	Mineral medium, sunlight

6.2. SELECTION OF ORGANISM FOR SCP

Various potential strains and substrates that could be utilized for SCP production are described. Nutritive value and removal of nucleic acids and toxins from SCP as a protein-supplementing source are discussed. New processes need to be exploited to improve yield. The (i) substrates utilized for growth, (ii) cultural conditions, (iii) treatment of cells and (iv) nutritional value are considered for selecting an organism for mass production of SCP.

i) **Substrates utilized:** Algae are photoautotrophic in nutrition, both eukaryotic algae and prokaryotic algae (cyanobacteria) utilize light energy with CO_2 as their source of carbon. Other groups of SCP microbes are mostly chemoheterotrophs, require complex organic nutrients. Agricultural residues such as straw, lignocellulose, starch and agro-industrial residues like, molasses, bagasse, milk whey, sulphite liquor are used as substrate for bacteria, fungi and yeasts.

ii) **Cultural conditions for mass production:** The pH of the medium is generally made acid for yeasts (4.5-5.5), neutral for bacteria (6.0- 7.5) and alkaline for algae (8.0-11.0). Blue green algae, *Spirulina* grows optimally in pH range of 9-11 and there is least chance of contamination of other microbes. The production of algae is feasible only in regions where there is bright sunlight and ambient temperature (25-35°C).

iii) **Treatment of cells:** Treatment of cells is directed towards removal of cell wall and reduction of nucleic acid content of microbes to make them suitable for human consumption. The low digestibility of SCP is connected with the cell wall fraction, which comprises 20% of the dry weight. *Spirulina* has a thin cell wall. Drying of cells and extraction of protein increases the digestibility of SCP. Microbial cells may contain 8 to 25 g nucleic acids per 100 g of protein. During evolution man has lost the ability to synthesize the uricase enzyme that digests the nucleic acid product, uric acid to allontoin. Accumulation of excess uric acid in blood leads to many pathological conditions. The concentration of nucleic acids is always below 5% of dry weight for *Spirulina.* Microbial cells have disagreeable flavour (yeasty odour) and unacceptable colour. Pre-treatment and processing of cells is required to remove disagreeable flavour and unacceptable colour. The SCP cells should be tested for their pathogenicity and toxicity. Toxic testing for gastrointestinal disorders, allergies and carcinogenic effects are carried out by feeding trails.

iv) **Nutritional value:** The high amount of protein in *Spirulina* is unusual even in the microbial world, being surpassed only by species of *Cellulomonas,* which may attain a protein level of 80%. In bacteria and yeasts however the high protein content is accompanied by a high content of nucleic acids that may exceed 25% of the dry weight. Egg albumin is a well balanced source of essential amino acids for human nutrition and the spirulina protein compares well except that it contains less of sulphur containing amino acids, methionine and cystine (Table 6.2).

TABLE 6.2 ESSENTIAL AMINO ACID CONTENT OF FOODS (g per 100 g Protein)

AMINO ACID	WHEAT	EGG WHITE	*SPIRULINA*	YEAST
Lysine	2.8	6.5	4.6	7.8
Threonine	2.9	5.1	4.6	5.4
Methionine	1.5	3.2	1.4	1.6
Cystine	2.5	2.4	0.4	0.9
Tryptophan	1.1	1.6	1.4	1.3
Isoleucine	3.3	6.7	6.0	5.3
Leucine	6.7	8.9	8.0	7.8
Valine	4.4	7.3	6.5	5.8
Phenylalanine	4.5	5.8	5.0	4.8

The digestibility, which is the percentage of amino nitrogen absorbed from intestine, is 70-90% for spirulina, the Biological value which is the percentage of amino nitrogen retained by the body is 30-60% and protein efficiency ratio (PER) which is the proportion of amino nitrogen retained when *Spirulina* is fed compared with that retained when egg albumin (reference protein) in fed, is 0.7-1.0. The protein and nutrition values of *Spirulina* are superior to other SCP. Microbes are generally rich in B group of vitamins. *Spirulina* contains carotene (a precursor of vitamin A) and tocopherol (vitamin E) in addition to B group of vitamins (Table 6.3).

TABLE 6.3 VITAMIN CONTENT OF *SPIRULINA*

VITAMIN	AMOUNT (mg per 100 g)
β CAROTENE	160
THIAMINE- B1	0.12
RIBOFLAVIN–B2	1.78
COBALMINE–B12	0.10
BIOTIN	0.22
TOCOPHEROL–E	**0.773 I.U**

The cultivation of algae for SCP production is seems to be more easy and economical than bacteria, fungi and yeasts. The substrates utilized for growth, cultural conditions for mass production and treatment of cells are simple for algae like *Spirulina*. Among the various microorganisms used as sources of SCP, the alga, *Spirulina* is considered as the best source. The composition of the biomass, including the high protein content, low content in nucleic acids, occurrence of high concentrations of vitamins and other growth factors and the presence of cell wall that is more easily digestible than that of other microbes indicate that SPIRULINA is a promising source of food or feed.

6.3. SPIRULINA SCP

Algae, as source of SCP for human consumption has been receiving worldwide attention. The blue green algae, *Spirulina* is rich in organic nitrogenous constituents. Species of *Spirulina* (Fig. 6.1) are multicellular, filamentous, non-heterocystous, non-nitrogen fixing cyanobacteria characterized by spiral structure. They have many spirals, appear as green filaments composed of cylindrical cells, arranged in un-branched helicoidal trichomes and a length of 300 microns (0.3 mm). The photosynthetic activity in *Spirulina* is carried

Fig. 6.1 *Spirulina* cells

out mainly in biliproteins-phycocyanin and allo phycocyanin. Phycocyanin is found in the concentration of about 10 per cent in *Spirulina*. They have been reported from quite different environments: fresh water streams, brackish water, sea, tidal ponds and saline ponds. *Spirulina* contains around 60% protein, 6% crude lipids, 15% carbohydrates, 0.1% β carotene and can play an important role in human nutrition as a protein supplement (Spirulina analysis Table 6.4).

TABLE 6.4 TYPICAL ANALYSIS OF SPIRULINA

PROTEIN	60.00% – 69.00%
CARBOHYDRATES	15.00% – 20.00%
LIPIDS	5.00% – 6.00%
MINERALS	7.00% – 10.00%
PHYCOCYANIN	12.00% – 20.00%
CHLOROPHYLL	1.00% – 1.50%
CAROTENOIDS	0.35% – 0.40%
BETA CAROTENE	0.15% – 0.25%
MOISTURE	4.00% – 6.00%
BULK DENSITY	0.55 – 0.65 g / cc
COLOUR	DARK BLUE GREEN

HISTORY: *Spirulina* is used for food from time immemorial by tribes living around Chad lake in Africa. The predominant species of phytoplankton of the lake is *Spirulina platensis*. The dried mat of *Spirulina* in the form of biscuits is used to prepare the sauce that accompanies millet meal eaten by the kanenbou population, the predominant ethnic group of Chad Republic. The potentiality of *Spirulina* as a food for human was exploited in ancient Mexico as well.

Spirulina maxima that grows abundantly in very alkaline waters of lake Texcoco near Mexico city was sun dried and used as food by the native Aztec population of Mexican valley at the time of Spanish conquest. *Spirulina maxima* grows practically as a monoculture for ten months per year in outer portions of lake Texcoco. It grows almost by itself on lakes and ponds and does not require any special care. Thus the two populations at considerable distance in time and space discovered independently the nutritional quality of *Spirulina*.

SPECIES OF *SPIRULINA*

Spirulina maxima, Spirulina platensis, Spirulina fusiformis

MASS PRODUCTION OF SPIRULINA

Spirulina is technologically amenable for large scale production. *Spirulina* production plants for mass cultivation are to be located in areas with suitable climatic conditions, particularly sunshine throughout the year. (Table 6.5) The semi-natural lake system provides an ideal environment for natural growth of *Spirulina*. The product is of low quality due to contamination and pollution in lakes and used for fish or animal feed.

TABLE 6.5 MAJOR PRODUCERS OF SPIRULINA

PRODUCER	QUANTITY- TONS PER YEAR
EARTH RISE SPIRULINA FARM,CALIFORNIA	150
TAIWAN, NANPAO SPIRULINA FARM	200
SIAM ALGAE, THAILAND	150
MCRC, CHENNAI, INDIA	300
CYANOTECH, USA	300
SOSA TEXCOCO, MEXICO	300

In artificially built cultivation system, waste water or clean water can be used for mass cultivation. The diluted sewage, liquid effluents from agro-industries or biogas plants are used after removal of solids and incorporation of nutrients. Fertilized sea water is also used in waste water system. The microalgae thus produced can be used as feed in aquaculture, animal feed and source of fine chemicals.

In clean water system, low cost medium with nutrients (urea, magnesium sulphate, sodium bicarbonate, rock phosphate and farm yard manure extract) promotes the growth of *Spirulina*. This system is more expensive because of construction of artificial cultivation farms. They have race way cement ponds with high quality nutrients and paddle wheels for air circulation. The *Spirulina* grown in clean waters under strictly controlled conditions could be used for human nutrition. Bureau of Indian Standard (BIS) has approved the food grade specifications for *Spirulina* in Standard No.IS 12895 amended in June, 1991.

GROWTH REQUIREMENTS FOR MASSCULTIVATION

Generally, circular cement tanks with shallow depth (25 cm) are constructed in required size according to convenience and yield needed. Open tanks are suitable for tropical and subtropical regions. The tanks are designed as race way ponds (Fig 6.2) with length/width ratio of 40. *Spirulina* requires effective pond management for outdoor production. Agitation of culture by paddles, rotators or pipe pumps provides aeration and efficient mixing of nutrients. Supply of CO_2 during day time improves the quality and yield. Maintenance of optimum levels of nutrients is mandatory, culture medium contains $NaHCO_3$, NPK, $MgSO_4$, at the ratio of 8: 1: 0.1. The depth of culture medium is maintained at 20 cm.

ALGAL TANK : RACEWAY POND

Fig. 6.2 Raceway pond

Spirulina grows well between pH 9 and 11 and there is a least chance of contamination at this pH. The optimum pH of nutrient medium is shifted from 8.4 to 9.5 during mass cultivation due to consumption of $NaHCO_3$. Atmospheric temperature of 30-35°C is ideally suited for getting higher yield of spirulina. The optimum light intensity for growth is between 20-30 kilo lux. The tropical environment in India is highly suitable for cultivation of *Spirulina*.

The alga is harvested by filtration. The culture liquid is pumped and filtered through nylon cloth in a rotary machine for particulate matter. The culture liquid is continuously recycled to culture ponds to become the nutrient base for the next crop. The drained Spirulina is wrapped in clean muslin cloth and pressed under weight of 50 kg to further drain moisture. The drained biomass is squeezed in the form of noodles on a dry clean cloth under sun. The sundried algal noodles are ground in a machine into powder. The algal dry powder is packed in small airtight plastic covers and ready for consumption. The yield is about 15 $g/m^2/day$ or 2-3 $Kg/m^2/year$ (20 tons/ha/year). The collected algal suspension is dried (sun dried or spray dried) and used as single cell protein.

Uses of Spirulina

1. Spirulina is a rich source of protein, minerals, and vitamins, crude fibres, used as supplement food in diets of undernourished poor children in developing countries.

2. Nutrients in *Spirulina*: High protein content (65-70%) with full spectrum of essential amino acids. Rich in vitamins β-carotene (precursor of vitamin A), vitamin E (tocopherol), B12 (cobalmine) B vitamins and niacin. Loaded with naturally chelated minerals calcium, phosphorus, iron. Contain essential fatty acid gamma linolenic acid and antioxidants.

3. *Spirulina* is very popular as a health food. It is low in fat and carbohydrate and has no cholesterol. Recommended for dieting and weight management of obese patients. It provides all the essential nutrients without excess calories and good for those who want to control obesity.

4. *Spirulina* possesses many medicinal properties. It lowers sugar level in blood of diabetics. It decreased the total serum cholesterol and low density lipoprotein cholesterol, while increasing high density lipoprotein. It is a good source of β-carotene, therefore helps in monitoring healthy eyes and skin. Spirulina rich in antioxidants

(β-carotene and tocopherol) is recommended for cancer treatment. It is an effective immunomodulator, an agent that can effect the behavior of immune cells.

5. Phycocyanin found in *Spirulina* is used as natural colourant in food, drug and cosmetics.
6. It possesses anti-oxidant, anti inflammatory and hepatoprotective properties.
7. *Spirulina* is widely used in beauty products such as face creams, hair lotions, high protein shampoo, face masks and biolipstics.
8. *Spirulina* contain high concentrations of carotenoids and as animal feed enhanced the pigmentation of eggs and meat.
9. Enzymes from *Spirulina* are used in molecular biology research, Phosphoglucokinase (PGK) for ATP determination and restriction enzymes in genetic engineering
10. Growing *Spirulina* in waste water is a method of recycling waste water as well as producing protein.

6.4. YEAST SCP PRODUCTION

Yeasts are rapidly multiplying microbes capable of utilizing readily available organic materials as their carbon source and energy substrate to convert cheap inorganic nitrogen compounds such as ammonium salts into nutritionally valuable cellular proteins. Many types of yeast are unable to breakdown starch and not suitable for food protein production. However, *Candida utilis* can be grown on starchy waste with *Endomosis fibuliger* that produces amylase and hydrocarbons are utilized as substrate by many types of yeast. Aerobically grown *Saccharomyces cerevisiae* utilizes molasses and produce a biomass of useful protein content.

BACTERIAL SCP PRODUCTION

PRUTEEN PROCESS: This process was developed by ICI utilizing *Methylophilus methlyotrophus* grown continuously on a sterile medium containing methanol, inorganic salts and ammonia. The pressure cycle fermentor employed has several unique features. The large quantity of air supplied is used to produce rapid internal circulation and thus no stirring mechanism is needed. The height of the fermentor creates a high pressure region at the bottom allowing rapid dissolving of oxygen and a low pressure region at the top, which causes release of carbon-dioxide. A suspension containing 8% dry weight of bacterial cells is continuously removed, concentrated and dried. The product contains 72% protein with an amino acid profile high in lysine and methionine.

MYCOPROTEIN PRODUCTION

A filamentous soil fungus *Fusarium graminarum* is mass produced for mycoprotein production. The organism is grown under steady state conditions for a period of several weeks in an aerobic fermentor. Hydrolyzed starch from potato, wheat, corn or molasses with inorganic mineral salts, sulphuric acid and ammonia gas is used as substrate. The medium is inoculated, fermented and biomass is separated in a continuous cyclone separator. The filter cake called mycoprotein is stored at 4°C for 48 hours. The product may be spray dried for long term storage at room temperature.

6.5. SINGLE CELL OIL (SCO)

Many microorganisms, *Fusarium pallidoroseum, F. lycopersici, Aspergillus sydowii* produce 45 to 50% oil from simple carbon sources. SCO organisms are potential for major modification in the fatty acid composition of their oils through mutagenesis and selection or recombinant techniques. The more directed recombinant techniques offer the possibility of designing the most appropriate oil for a particular food (docosa hexenoic acid - DHA, containing oil for infant formula) or clinical application (poly unsaturated fatty acid - PUFA, containing oil for heart patients). Oleagenous yeast, *Candida curvatum*, is a potential source of SCO. Selected clones of *Candida curvatum* with depressed levels of desturase activity produce oil rich in stearate and the oil may have much higher value as a cocoa butter substitute. As a competitive product for cocoa, the SCO become much more commercially attractive.

6.6. FOOD CHEMICALS

A variety of food chemicals and additives may be produced by microbial fermentation and used in food processing (e.g. organic acids: citric acid, acetic acid, lactic acid, enzymes: amylase, protease, pectinase, microbial rennet, polysaccharide : xanthan, food colourant: β-carotene) Table 6.6.

TABLE 6.6 FOOD CHEMICALS PRODUCED BY MICROORGANISMS

FOOD CHEMICAL	ORGANISM	APPLICATION
Glutamic acid	*Corynebacterium, Brevibacterium*	Monosodium glutamate used as flavour enhancer
Lactic acid	*Lactobacillus delbrueckii*	To acidify jams, jellies, confectionery, soft drinks
Citric acid	*Aspergillus niger*	Added to soft drinks and candies to reduce the pH
Vinegar	*Acetobacter, Gluconobacter*	Condiment, preservative
Dextran	*Leuconostoc mesenteroides*	Stabilizer in ice creams, syrups and confections
Xanthan	*Xanthomonas campestris*	Stabilizer, not digested by animals and humans
Astaxanthin	*Xanthophyllomyces dendrorhous*	Colouring agent in food
β Carotene	*Dunaliella salina*	Colouring agent in food
Vanillin	*Bacillus subtilis*	Fragrance agent in food
Red Food colourant	*Monascus purpureus, M. ruber*	Red mold rice-'Angkak'

QUESTIONS

1. Define Single cell protein (SCP).
2. What are the advantages and disadvantages of SCP?
3. List out the various microbes used as SCP.
4. What are the criteria considered for selecting an organism for Mass production of SCP?
5. write a brief account on treatment of SCP cells.
6. Explain the nutritional value of *Spirulina*.
7. Describe the mass cultivation methods and uses of *Spirulina* SCP.
8. Writ short notes on: Mycoprotein, Single cell oil, PRUTEEN PROCESS.
9. List out food chemicals produced by microbes and their applications.
10. Which of the following is not a character of *Spirulina*?

    a) *Spirulina* grows well between pH 8.5 to 9.5.

b) *Spirulina* is rich in beta carotene

c) *Spirulina* is grown in medium rich in organic nutrients

d) *Spirulina* is a cyanobacterium.

11. Where *Spirulina* is grown for mass production?

a) Trays with solid substrate b) Anaerobic fermentor

c) Race way pond d) Paddy straw bags

7

MUSHROOM CULTIVATION

Mushrooms are most ancient, best known and greatly admired fleshy macrofungi. Mushrooms are delicacies, much sought after and greatly valued foods. Wild mushrooms are mostly found in forests, tree trunks, pastures, lawns and road side during winter season and appear at surprising speed. Mushroom hunting or gathering is a popular hobby in Western countries. The wild mushrooms may be edible or unpalatable or poisonous. Poisoning by wild mushrooms is common and may be fatal or produce merely mild gastrointestinal disturbance or slight allergic reaction. The term toad stool is popularly reserved for inedible or poisonous mushrooms, but this classification has no scientific basis. This alarming term keeps many people away from these foods. It is important that every mushroom intended for eating be accurately identified. Mushrooms vary in colour, size, shape, growth habit, yielding ability and other characteristics.

STRUCTURE: Mushrooms are the fleshy fungi which constitute a major group of lower plant kingdom. Mushroom is the conspicuous umbrella-shaped fruiting body (sporophore) of certain fungi of the order *Agaricales* in the class Basidiomycetes but also of some other groups. Popularly, the term mushroom is used to identify the edible sporophores. The edible reproductive part consists of a stem and a cap. Gills found on the underside of the cap, are arranged like spokes in a wheel. The gills bear multiple spores which start new fungi. The sporophore emerges from an extensive underground network of fine thread like white growth known as mycelium. The mushroom, fruit body produced basidiospores at the tip of club like structures, called basidia, which are arranged along the gills of the mushroom. Beneath the mushroom, in the soil, is the mold colony itself, consisting of a mat of intercoiled hyphae, sometimes several feet in diameter. The mushrooms first appear as white tiny ball consisting of short stem (stipe) and a cap (pileus) which begins to open up like an umbrella. The delicate membrane or Veil (velum) enveloping the cap tears off, if allowed to develop fully, and lamellae (gills) radiating from the stalk into the cap become visible. These gills become darkened as the basidiospores (seeds) develop into millions and fall in the ground for starting their life cycle once again for second generation of mushrooms.

Mushroom mycelium is extremely durable surviving for long periods of dryness. As long as nourishment is available and temperature and moisture are suitable, a mycelium will produce a new crop of sporophores each year during its fruiting season. As soon as warm damp conditions set in, the fruiting bodies develop. At first these are tiny roundish buttons which pierce through the leaf litter above the ground. They then grow a stalk and gain height, a few centimeters to several inches, depending upon species. This entire emergence of a mushroom up into view can be overnight. In a very restricted sense, mushroom indicates the common edible fungus of fields and meadows (*Agaricus campestris*). A very closely related species, *A. bisporus,* is the mushroom grown commercially and seen in markets.

EDIBLE MUSHROOMS

Edible mushrooms are treated as special food from early days of civilization. The use of edible mushrooms for food dates back to atleast to early Roman times. Mushroom cultivation started in France during 17th century. The cultivated mushroom is a horticultural adaptation of common field mushroom that is found in pastures and grassy places. They are popular for their delicacy, flavour, fine texture, nutrients and favourable health benefits. Mushrooms farming is becoming successful because of its very low inputs. It is estimated that about 300 million tons of fresh mushroom can be produced from just one fourth of world's annual yield of straw (2,353 million tons). It was calculated that approximately 317 million metric tons of fresh mushroom could be produced annually that would provide 197 g of fresh mushroom daily to each person in the world.

Since mushrooms grow independent of sunlight, they can be grown in complete darkness, but the darkness is not an essential pre-requisite. They are relatively fast growing, do not require fertile soil since grown on composted agro wastes and their culture can be concentrated within a relatively small space. In addition to floor, air space is also utilized resulting in higher production.

The worldwide annual production of edible mushroom is approaching three million tons (Table 7.1). In India, it has added importance in view of the vegetarian habits of a considerable population. Among the edible species only cultivated ones can provide a stable supply. About 2000 species of fleshy fungi are regarded as prime edible mushrooms but only about 40 of them economically cultivated, around 20 commercially cultivated and only the following six have reached the industrial scale.

TABLE 7.1 ANNUAL PRODUCTION OF EDIBLE MUSHROOM (2000)

COMMON NAME	SPECIES	PRODUCTION (TONS)
White button mushroom	*Agaricus bisporus*	1300,000
Shiitake mushroom	*Lentinus edodes*	350,000
Paddy straw mushroom	*Volvariella volvacea*	200,000
Oyster mushroom	*Pleurotus* spp	180,000
Wood ear mushroom	*Auricularia* spp	130,000
Winter mushroom	*Flammulina velutipes*	110,000

ADVANTAGES OF MUSHROOMS

1. Mushrooms are grown in soil-less culture and no need for arable land and sunlight.
2. Mushrooms are good sources of essential amino acids, vitamins of B complex, folic acid, available iron, potassium and phosphorus.
3. Mushroom cultivation produces more protein per unit area of land and time than any other form of agriculture.
4. Mushrooms are low calorie foods, as starch is absent, good for diabetic patients. The rich quality dietary fibre is favoured against gastrointestinal ulcers. They are demonstrated to possess hypocholesterolemic, hypolipidemic, antitumorous, anti-bacterial and antiviral activities.
5. Mushroom technology does not involve the use of any chemical or pesticide and aptly regarded as Organic food.
6. The cultivation of edible mushrooms in agricultural residues, the wastes are well utilized, recycled and environment is not endangered of pollution.
7. Mushrooms have medicinal properties. *Lentinus, Coriolus, Schizophyllum* and *Ganoderma* species of mushrooms have medicinal properties like immune enhancers, anti cancerous, diet remedy for diabetics, etc. The concept of Selenium rich mushrooms for slowing down the progress of AIDS has been gaining importance in the

recent years. Some mushroom extracts induce formation of interferon, a defense mechanism against viral infections.

NUTRITIVE VALUE

Mushrooms have long been known as "Queen of vegetables", as they are rich in protein, vitamins and minerals. They contain more protein than any other vegetable and it is easily digestible. By fresh weight, the common commercially grown mushroom is more than 90 percent water, less than 3 percent protein, less than 5 percent carbohydrate, less than 1 percent fat, and about 1 percent mineral salts and vitamins (wet weight) (Table 7.2).

TABLE 7.2 NUTRITIVE VALUE OF COMMON EDIBLE MUSHROOMS

NUTRIENT CONTENT	BUTTON MUSHROOM	STRAW MUSHROOM	OYSTER MUSHROOM
PROTEIN (Dry Wt)	26.0 %	22.0%	22.0 %
CALORIES (Kcal/100g Dry Wt)	360	370	360
FAT	2-8 %	6-7%	1-2%
VITAMINS (mg/100g Dry Wt)			
Thiamine	1.0- 9.0	1.0-2.0	5.0
Riboflavin	4.0-5.0	3.0-4.0	5.0
Niacin	43.0-57.0	90	110
Ascorbic acid	20.0-80.0	20	10

Mushrooms are considered as health food because their mineral content is higher than any vegetables. Furthermore, protein content of fresh mushroom is about twice that of vegetables. Mushrooms protein contains all nine amino acids-isoleucine, leucine, lysine, methionine, phenylalanine, threonine, valine, tyrosine, tryptophan - essential to the diet of humans. Mushrooms are rich especially in leucine and lysine, which are lacking in most staple cereal foods. With respect to essential amino acid indices, the nutritional value of mushroom almost equal to that of milk. Mushrooms also contain many vitamins such as thiamine, riboflavin, niacin, ascorbic acid and devoid of starch and calories. Mushrooms therefore, are ideal food for diabetics and any one counting calories. Although mushrooms contain protein, vitamins and minerals, they are largely composed of water (90%) and hence of limited nutritive value. The overall nutritive value of mushrooms is somewhere between vegetables and meat.

MUSHROOM CULTIVATION IN INDIA

Mushroom cultivation started in India around 1900. In 1950s Paddy straw mushroom cultivation was started in Coimbatore and in 60s oyster mushroom was grown in Mysore. In Solan, Himachal Pradesh, synthetic compost containing horse dung and wheat straw was developed for growing Button mushroom. In 1974, a unit for bulk pasteurization of compost and casing soil was established in Solon to meet the requirements of mushroom growers of Himachal Pradesh. Mushroom cultivation was introduced in Kashmir valley in 1975. The State Agricultural Universities (SAU) established in India popularized the mushroom technology among the farmers and each SAU adapted few villages as mushroom villages. Central Food Technological Research Institute (CFTRI), Mysore, developed technology for production, processing and export of edible mushrooms (Button mushroom, Oyster mushroom, Paddy straw mushroom) as organic food. CFTRI introduced vertical bed system (urban and rural models) for large scale production of Oyster mushroom to suit industrial needs, to conserve space, time and labour.

7.1. BUTTON MUSHROOM CULTIVATION-*Agaricus bisporus*

There are three distinct operations in button mushroom cultivation, they are

i) Production of compost (substrate)

ii) Production of spawn (mycelium)

iii) Inoculation of compost with spawn for mushroom growth and yield.

i) **Production of compost (substrate):** Mushrooms can be grown successfully in a wide variety of inexpensive substrates such as agricultural and agro-industrial residues. Horse manure has long been used but other organic materials are suitable. In view of the scarcity of horse manure in some places alternative substances of vegetable origin suitable for mushroom production are also used. Some organic materials have been used alone or in combination with horse manure are corncobs, cornstalks, hay, straw, sawdust, dried brewers grains, poultry manure, banana leaves and cotton waste. The controlled production of composted substrate using mixtures of plant residues and manures or other nitrogen rich materials as activators stimulates mushroom growth. The purpose of compost preparation is to provide medium for rapid growth of mycelium.

The trays are filled with fresh compost substrate and usually 10 Kg compost is filled per sq. foot of six inches deep tray. The compost is allowed to ferment at 60°C for 4 to 6 days or until release of ammonia disappears. The compost could be pasteurized by steam generated by a boiler to kill the vegetative cells of microorganisms.

COMPOSITION OF COMPOST

Horse manure	-	1000 Kg
Wheat straw	-	500 Kg
Chicken manure	-	300 Kg
Brewers grain	-	60 Kg
Gypsum	-	30 Kg
Urea	-	7 Kg

ii) **Production of spawn:** Spawn is the propagating material for mushrooms. It contains spores mixed with spawning materials such as manure, tobacco stems and cereal grains. The spores collected from matured mushroom fruiting body are collected under aseptic conditions and inoculated into cooked cereal grains in a bottle. The spores germinate and cover the grains with mycelial growth. Spawn is most often purchased from specialty spawn makers.

iii) **Inoculation of compost with spawn for mushroom growth and yield:** The compost bed temperature is allowed to fall below a level before the spawn is introduced into the bed as inoculums, because high temperature damages the spawn. Spawning is carried out when the compost reaches a temperature of 25°C. The spawn is inoculated either by broadcast method or by planting small pieces, two inches deep in compost tray. After spawning the trays are held in spawn run room at a temperature of 25°C and a relative humidity of 80-90%. The growth from the mushroom spawn spreads through the upper three inches of the bed of compost. Good mycelial growth is essential for button mushroom (Fig. 7.1) production.

Fig. 7.1 Button Mushroom

The practice of placing a thin layer of soil or peat on top of the compost, after the spawn has run through it is called **casing.** After two weeks of growth, a one inch layer of casing soil is spread over the surface of the bed. Fruiting body initiation is induced by this process. Microbial activity in this layer is responsible for the initiation of fruiting combined with lowering of carbon dioxide concentration in the ambient air surrounding the cultures. After casing, trays should be maintained at 10 to 16°C and kept sprinkled with water until the first tiny mushroom appears in about 3-4 weeks. After that, watering is done only as needed. The mature fruiting bodies appear in a series of cycles (flushes) over intervals of 7-10 days. Picking may commence at the first sign of buttons, and last for 6-7 weeks. The first two flushes have the greatest biomass and then yield decreases at each subsequent flush.

7.2. SHIITAKE MUSHROOMS-*Lentinus edodes*

This is the major edible mushroom grown in Japan and it is estimated that there are 250,000 growers in Japan. This wood destroying fungus produces attractive brown-red fruit bodies and sold as both as the fresh and dried product. Shiitakes are grown by inoculating, mushroom spawn into holes drilled into suitable species of hardwood logs. It grows on the wood of broad leaved trees, mainly of the family Fagaceae. Species of oak have proven most satisfactory in U.S. These trees are normally felled in autumn, when the sugar content of the wood is high after leaf fall. The felled timber, generally 5-15 cm in diameter, is cut into 1 meter length and stored for 1-2 months before inoculation. The holes in logs are made using specially designed wedge shaped axes or power drills. Usually 15-20 inoculation points are made in each bed log. After inoculation the logs are placed in a laying yard in obliquely upright position. The optimum temperature for wood colonization is between 24 and 28°C. Completion of colonization takes up to one year. The fruit bodies start appearing from second year onwards. The mushrooms are rich in vitamins and also contain number of substances effective against blood cholesterol, anti tumor principle and substances inhibitory to the development of AIDS (lentinan sulphate).

7.3. CULTIVATION OF PADDY STRAW MUSHROOM-*Volvariella volvacea*

Paddy straw mushroom is known as a delicacy for many centuries in China and other Asian countries. It is mainly grown in paddy straw, hence the name Paddy straw mushroom. These fungi derive their carbonaceous food by decomposing lignin, cellulose and hemicellulose present in agro-wastes with the help of extracellular enzymes secreted by the mycelium.

In poly-bag method of Paddy straw mushroom cultivation the following steps are involved

1. Chop the fresh dried paddy straw into 1-2 inches bits and the bits are soaked in water for 24 hours or autoclaved.
2. Drain the excess water from the Paddy straw bits
3. Add horse gram powder at 8 g per Kg of straw and mix them thoroughly
4. Pack the polythene bags with straw-gram powder mixture in layers and add spawn in between the layers at 30 g per Kg of straw.
5. Puncture the bags with needles to facilitate exchange of air and draining of excess water.
6. Tie the mouth of the filled bags and incubate the bags for two weeks in a room at 30-35°C with sufficient light and ventilation for spawn running.
7. After two weeks, cut open and remove the polythene bag on the sides without disturbing the straw mixture.
8. Maintain the humidity of the contents to about 85-90% by spraying water.
9. After 3-4 days of opening of the polythene bags, small pins appear on the substrate.
10. First harvest is obtained 3-4 weeks after spawning and subsequent harvests could be made in a week gap.

7.4. CULTIVATION OF OYSTER MUSHROOM-*Pleurotus* spp.

Pleurotus spp. are commonly known as Oyster mushroom (**Fig. 7.2**). Unlike other mushroom species, *Pleurotus* spp are the easiest, fastest and cheapest to grow, require less preparation time and production technology. Also the first flush is usually large, without the need for compost, manure, limestone, casing or temperature shocks. With more than 100% biological efficiency, coupled with distinct flavour, aroma and excellent drying and preservation qualities it is assured a unique status as a delicacy. The oyster mushroom has a pleasant,oyster like flavour and is often prepared by dipping in egg and frying slowly. This mushroom grows in bracket-like clusters on

Fig. 7.2 Oyster mushroom

decaying tree trunks. It is almost stem less. The fleshy, tender cap is 8 to 13 cm across, tawny olive-colored when young, but fades with age.

Bioconversion of ligino-celluosic residues through cultivation of *P. sajor-caju* offers the opportunity to utilize renewable resources in the production of edible, protein rich food that will sustain food security for people in developing countries. Cultivation of edible mushrooms is one of the most economically viable processes for the bioconversion of ligino-celluosic wastes. The technology can also limit air pollution associated with burning agriculture wastes as well as to decrease rodent, pests and deleterious fungal inoculum populations.

Pleurotus spp. are wood destroying saprophytic fungi colonize rapidly on unfermented natural liginocellulosic wastes. Because of their shape they are called as Oyster mushroom. *Pleurotus spp.* can be cultivated in plains throughout the year, both during cooler months and summer. In temperate countries *P. ostreatus* is the commonly cultivated species. In tropical countries like India, species *P. sajor-caju, P. citinopileatus, P. flabellatus, P. platypus, P.sapidus* and *P. florida* are commonly cultivated in shredded straw substrate supplemented with nitrogenous substances.

Various agricultural by-products are being used as substrates for the cultivation of Oyster mushrooms. Some of these wastes include banana pseudostem, banana leaves, peanut hull, corn straw, mango fruits and seeds, sugarcane leaves, wheat and rice straw. Coffee pulp, coir waste and cotton waste are other substrates used. The widely used substrate for cultivation of oyster mushroom in Asia is rice/paddy straw. It is also considered the best substrates in terms of yield and high protein content. In Europe, wheat straw is used, while in South

East Asian countries saw dust is more common. In gulf countries palm is used for cultivation of oyster mushrooms since date palm results in the accumulation of large quantity of lignin and cellulose and are available without any cost throughout the year.

Grain spawn is used for inoculation. The substrate is packed in polythene bags and kept in dark room at 25-35°C with 80% humidity. The complete colonization of the substrate is achieved in 2-3 weeks. The fruiting is light dependent. Fruit bodies start developing in about 2 weeks after complete spawn running. In 25- 35 days, two to three flushes of fruit bodies can be harvested. The yield is around 1.0-1.15 kg fresh weight per Kg dry substrate. The bioconversion efficiency is recorded up to 115%.

COMPARISON OF COMMON EDIBLE MUSHROOM CULTIVATION

The common edible mushrooms-button mushroom, shiitake mushroom, paddy straw mushroom and oyster mushroom are compared for their nature of substrates, temperature required, production time, yield in grams fresh per Kg of dry substrate, protein conversion efficiency and uses of spent (Table 7.3).

TABLE 7.3 COMPARISON OF COMMON EDIBLE MUSHROOMS

ASPECTS	BUTTON MUSHROOM	SHIITAKE MUSHROOM	STRAW MUSHROOM	OYSTER MUSHROOM
Nature of substrate	Compost	Wood logs	Rice straw	Lignocellulose
Temperature required	< 20°C	12 -20 °C	32-36°C	15-30°C
Production time	60-90 days	105 days	20-30 days	25-35 days
Yield in g fresh wt per Kg dry substrate	700	250	300-400	1000-1150
Protein conversion efficiency	18.5	-	11.5	21.7
Uses of spent	as casing and manure	as manure	as manure	substrate material for biogas and paper manufacture

QUESTIONS

1. Write on the significance of mushrooms
2. Comment on nutritional value of edible mushrooms.

3. List out various edible mushrooms and their substrates
4. What are the advantages of mushroom cultivation?
5. What are the basic operations in Button mushroom cultivation?
6. Describe the shiitake mushroom cultivation
7. Describe the steps in Paddy straw mushroom cultivation
8. Describe cultivation of oyster mushroom
9. Compare the cultivation of common edible mushrooms
10. Write short notes on Shiitake mushroom, *Pleurotus* spp., Production of spawn.
11. The spawn is prepared from fungal

    a) Hyphae b) Spores

    c) Rhizoids d) Enzymes
12. This is used as substrate for spawn preparation

    a) Cereal grains b) Coir pith

    c) Sea water d) Pebbles
13. Paddy straw mushrooms are cultivated by this method

    a) Polybag method b) Wood log method

    c) Compost tray method d) All the above
14. The major component of button mushroom compost is

    a) Straw b) Horse manure

    c) Limestone d) Urea

8

MICROBIAL ENZYMES

Several attempts have been made to detect the biopotentials of microorganisms with reference to their ability to produce various enzymes with a view to use them in industrial processes and bio-remediation applications. Enzymes are increasingly used as a new route to synthesize food ingredients such as flavours, emulsifiers, sweeteners, food additives, nutrients. With the advent of genetic engineering, a wide range of enzymes has become available on a large scale and this has increased the scope of enzyme technology.

MICROBIAL ENZYMES: Enzymes are proteinaceous complex organic catalysts of biological origin and vital constituents of cells. Microbes produce a variety of enzymes, intracellular and extracellular enzymes. Most of the microbial enzymes produced for commercial usage are extra cellular enzymes. Extracellular enzymes are secreted outside the cell and hydrolyze polymeric substances. Amylases and proteases are enzymes of economic importance. At present over 60 enzymes are in commercial use, of which three-quarters are hydrolytic enzymes. Commercial production of microbial enzymes utilizes a variety of fungi, yeasts and bacteria. The composition of the medium, genetic stability of the microbial strain and physical and chemical factors of fermentation influence the production of a particular enzyme.

Enzymes catalyze a variety of reactions during processing and preservation of food products and form a large group of processing aids. Food industry utilizes enzymes in the production of bakery items, dairy products, fruits and vegetables juices and products, tenderization of meat, modification of oils and fats, flavour production and preservation of food products. Enzymes have also found use in the manufacture of ingredients like glucose, high fructose syrup, citric acid, colour and flavour constituents.

8.1. AMYLASES

INDUSTRIAL PRODUCTION OF ALPHA AMYLASE

Fungal amylases for commercial purposes are derived from *Aspergillus oryzae.* The enzyme is also produced by *Aspergillus niger, Aspergillus awamori, Rhizopus* sp. The fungal amylases are produced in solid substrate culture with wheat bran as substrate. For this purpose wheat bran spread in relatively thin layers in trays is moistened with water or dilute acid, sterilized and inoculated with spores of *Aspergillus oryzae.* After the incubation time, which varies from 2 to 4 days, the fungus plus bran can be dried at 50°C and then ground to serve as amylase preparation. However more often the amylase is extracted by water from wheat bran culture, precipitated from the aqueous solution by the addition of alcohol and dried at 55°C. *Aspergillus niger* is suitable for amylase production under submerged aerated culture. A starch salt medium is used for fermentation and addition of corn steep liquor or yeast extract stimulates the production of the enzyme.

Bacterial α amylases obtained from *Bacillus licheniformis, Bacillus amyloliquefaciens* are used in food processing and food fermentation.

Applications

i) Amylases convert starch into sugars, syrups and dextrins.
ii) They are used in baking-The enzyme elevated the levels of fermentable monosaccharide and disaccharide of the dough to a concentration that activates yeast. The enzyme prevents the production of sticky crumbs in bread.
iii) Fungal amylases are used for saccharification of starch to produce dextrose or glucose and preparation of maltose syrup.
iv) In brewing and alcohol fermentation industry for starch hydrolysis and in starch process industries for desizing textiles amylases are used.
v) Beta amylase from *Clostridium* species is used to produce maltose rich syrup from corn starch.

8.2. PROTEASES

PRODUCTION AND USES: Proteases are industrially produced by species of *Aspergillus oryzae, Mucor and Rhizopus.* These protein splitting enzymes from microorganisms are widely used in protein industries, where the protein containing materials from plants and animals are processed into foods and feeds. This imparts improved textural, nutritional and sensory properties to the products and in addition making maximum utilization of proteins. Fungal protease is added to bread dough during fermentation stage. It releases amino acids and peptides necessary for yeast growth and multiplication. They improve handling characteristics of dough, increase dough volume and enhance the texture and elasticity of gluten.

Proteases are used to tenderize the tough meat as they digest both structural and connective proteins. They are used in brewing industries for chill proofing of beer. They are also used in detergents, leather industries (dehairing), pharmaceutical industries and manufacture of soya bean products.

8.3. GLUCOSE ENZYMES

Starch is a non-sweet, abundantly available carbohydrate, upon hydrolysis, either by enzyme or non-enzymatic methods, it yields glucose. Mainly starch is hydrolyzed by alpha amylase and amyloglucosidase obtained from microorganisms to produce glucose syrup from starch. High Fructose Corn Syrup (HFCS), is an important sweetener, used in confectionery industry. It is manufactured from corn steep liquor by using three different enzymes in a sequence, amylase for liquefaction of starch, followed by glucoamylase to produce glucose and finally glucose isomerase to convert glucose to fructose. FLOW CHART 8.1.

FLOW CHART 8.1 ENZYMATIC DERIVATION OF FRUCTOSE FROM STARCH

STARCH SLURRY
↓ GELATION LIQUEFACTION
ALPHA AMYLASE
MALTO DEXTRINS
↓ SACCHARIFICATION
GLUCOAMYLASE
GLUCOSE SYRUP
↓ ISOMERIZATION
GLUCOSE ISOMERASE
FRUCTOSE SYRUP
↓ REFINING
VERY HIGH FRUCTOSE SYRUP

Organism Producing Glucose isomerase: *Flavobacterium arborescens, Bacillus licheniformis, Arthrobacter* sp. *Streptomyces phacochromogens, Streptomyces nigrificans, Streptomyces olivaceous.*

Industrial Uses: It has been mentioned earlier the hydrolysis of starch began in early 1960's to prepare dextrose and glucose syrups. Currently various enzymatic processes are applied for various products. Glucose isomerase is an important enzyme used commercially in conversion of glucose to fructose syrup. The reaction mixture at the end contains 42% fructose, 52% glucose and 6% dextrins. The mixture is sweeter than glucose and as sweet as sucrose. New techniques have been developed to obtain 55% fructose concentration in syrup.

Glucose isomerase is used in pharmaceutical industry for the production of semi synthetic penicillin and fructose syrup respectively. Alpha amylase from *Bacillus subtilis, Aspergillus oryzae* or *Bacillus licheniformis* and beta amylase from malting cereals hydrolyze corn starch into a mixture of glucose, maltose and dextrins to give starch syrup. The enzyme glucoamylase hydrolyzes starch into D-glucose (saccharification).

8.4. PECTIC ENZYMES

All types of fruits and berries contain significant amounts of pectin, which is partially soluble. When the juice is extracted from the fruits, some of the pectin passes into juice of fruit, which becomes vicious and later settle down. In the manufacture of wine or apple and grape fruit juices, pectinases are added to pulp to hydrolyze the pectin, so that the juice becomes clear. Commercial pectic enzymes, which are referred as pectinases are a mixture of pectin esterase (EC 3.1.1.11) and polygalacturonase (EC 3.2.1.15).

Pectic enzymes are produced by species of *Aspergillus, Botrytis, Penicillium,* and *Rhizopus*. Pectinolytic enzymes can be produced either by submerged or solid state fermentation. Solid state fermentation is defined as growth of microorganisms on solid material in the absence of free water. It has tremendous potential in enzyme production because of the advantages such as simplicity, high productivity and concentrated products over submerged fermentation. Solid state fermentation is not suitable for automation.

The enzymes play a specialized role in citrus fruit processing. It improves the stability of the cloud, a desirable feature in citrus fruit juice. The desirable characteristics of colour and flavour are present in the insoluble cloud constituents. They are widely used in fruit juice extraction, fruit juice clarification and enzymatic maceration and liquefaction of fruit and vegetable tissues. They are used in the processing of green coffee beans to hasten the removal of the jelly that surrounds the coffee cherry and improve the quality.

CELLULASES

Cellulases are industrially produced from *Trichoderma viridae* and *Aspergillus niger.* Cellulases are used to improve palatability of low quality vegetables. They increase the flavour of mushrooms, promote extraction of natural products and alter the texture of foods. They are used in the extraction of green tea components, soybean, coconut cake, sweet potato, corn starch and agar. They are used for production of vinegar from citrus pulp and sea weed jelly. They are also used as digestive aids together with other hydrolytic enzymes.

INVERTASES

Invertase is widely used for hydrolysis of sucrose. The hydrolysis (inversion) of sucrose to glucose and fructose facilitate more sweetness, enhancement of fruit flavour, high stability at low pH and is less likely to crystallize in products. The yeast invertase produced by *Saccharomyces cerevisiae* is used in confectionery industry and to make syrup of invert sugars (glucose and fructose) and D-fructose from inulin. It is also used for hydrolysis of sucrose in molasses based microbial fermentations.

LIPASES

Cocoa butter is highly priced valuable food ingredient. A substitute having almost similar properties has been enzymatically synthesized from cheap palm oil. In this process triglycerides of palm oil are inter esterified replacing palmitoyl substitutents with steric acid by using lipase from *Rhizopus* species. Lipases are used for flavour development in processed cheese products, production of butter fat flavours in chocolates, flavour modification of cooking fats, improved whipping properties of egg white and antistaling action in bakery foods,

Long chain fatty acids have been reported to have important physiological role in humans. In addition they have beneficial role against cardiovascular disease and cancer. Generally vegetable oil does not contain these fatty acids. Using lipase enzymes, vegetable oils can be enriched with these long chain fatty acids.

MICROBIAL RENNETS

Rennets are proteolytic enzymes that are used for the coagulation of milk in cheese production. The most common rennet enzyme in commercial use throughout the world is calf rennet, an enzyme extract obtained from the fourth stomach (abomasums) of calves. However declining calf slaughter has resulted in a shortage of rennet extract. The major product of dairy industry is cheese. For the preparation of cheese, enzyme rennin is required which acts on milk protein and makes it to coagulate. The calf rennin, chymosin, is now produced by recombinant DNA technology in microbes is widely used in cheese manufacture to enhance ripening process and increase the cheese flavour production. Many fungal and bacterial cultures have the potential to produce a milk clotting enzyme with suitable milk clotting to proteolytic ratio. The microbes explored for potential include *Mucor pusillus, Mucor micheli, Endothia parasitica* as fungal sources and *Bacillus cereus, Bacillus polymyxa,Bacillus coagulans, Bacillus subtilis* as bacterial sources.

Wheat bran moistened with water is sterilized at 121°C for 20 minutes used as substrate for fungal rennet production. Inoculated medium is incubated at 22°C for four days. Distilled water is added to the mass culture, freeze at 0°C and thaws the contents at room temperature. Then filtered through muslin cloth and centrifuged at 5000 rpm in refrigerated centrifuge. The precipitate is dissolved in water for use.

LACTASE

Lactase or β galactosidase converts lactose in to glucose and galactose, is produced by *Aspergillus niger, A.oryzae, Kluyveromyces lactis, K. marxianus, K. fragilis, Lactobacillus amylophilus* used in the production of galacto-oligosaccharides and added to infant milk, prevent crystallization of lactose during food preservation, increase sweetness and increase the solubility of milk products.

PHYTASES

Phytases hydrolyze phytic acid in to myo-inositol and inorganic phosphates is produced by *Aerobacter aerogenes* and *Bacillus subtilis*. Phytic acid act as antinutrient factor, form complex with starch, protein, lipid and make them unavailable, inhibit digestive enzymes, chelate cations Calcium, Magnesium, Iron. Phytic acid in food ingredients, wheat flour, rice flour, gram flour and animal feed supplements are dephytized by phytase.

8.5. APPLICATIONS OF ENZYMES

Many exciting changes have been happening in food industries as a result of improvements and innovations in the use of enzymes to food materials. Better control of enzymes reactions in food, the utilization of enzymes and development of new enzyme uses are dependent on understanding of structure and mechanism of action of enzymes. The enzymes added to food function as preservatives, antioxidants, macerating aid, oil extracting agent, peeling and slicing aid. Industrial production of enzymes is needed for the commercial production of food and beverages (Table 8.1). Use of microbial enzymes in food processing falls roughly into three categories.

1. Those in which enzymes form an essential part of a process, e.g. production of cheese, beer, spirits.
2. Those in which enzymes are used to improve the economics of a process. e.g. extraction of fruit juices and essential oils
3. Those in which enzymes are used to improve product quality e.g. meat tenderization, bread loaf volumes, flexibility of glucose syrups and dextrins.

TABLE 8.1 APPLICATION OF MICROBIAL ENZYMES

MICROBIAL ENZYME	PRODUCING ORGANISM	APPLICATION
α Amylase	*Aspergillus oryzae, Bacillus amyloliquefaciens*	Preparation of glucose syrups Removal of starch size, Liquefaction of brewing adjuncts
β-Gluconase	*Aspergillus oryzae, Bacillus amyloliquefaciens*	Preparation of glucose syrups
Glucoamylase	*Aspergillus niger, Rhizopus* sp.	Starch hydrolysis
Glucose Isomerase	*Arthrobacter* sp.	High fructose corn syrup
Lipase	*Candida lypolytica Saccharomyces lipolytica*	Flavour in cheesevegetable oils can be enriched with long chain fatty acids
Pectinase	*Aspergillus* sp.	Clarification of wine, fruit juices
Protease-acid	*Aspergillus* sp.	Substitute for calf rennet
Protease-alkali	*Aspergillus oryzae, Bacillus* sp.	Dehairing of hides Detergent additive
Protease-neutral	*Bacillus amyloliquefaciens Bacillus thermoproteolytics*	Liquefaction of brewing adjuncts

Cellulase	*Trichoderma reesi*	Improve palatability of low quality vegetables, increases the flavour of mushrooms, promotes extraction of natural products
Invertase	*Saccharomyces cerevisiae*	Facilitate more sweetness, enhancement of fruit flavour
Microbial rennet	*Mucor pusillus, Mucor micheli*	Cheese production
Lactase	*Kluyvermyces fragilis*	Digestion of lactose
Phytase	*Aerobacter aerogenes*	Digestion of phytin and release of phytin phosphorus in food grains.

8.6. APPLICATION OF IMMOBILIZED ENZYMES

Immobilization of an enzyme has been defined as the "imprisonment of an enzyme in a distinct phase that allows exchange with, but is separated from, the bulk phase in which substrate, effector or inhibitor molecules are dispersed and monitored". Tailoring of enzymes for specific function is now available with the advent of enzyme immobilization methods. Application of immobilized enzymes in food industries is restricted to a very few cases. For instance, aminoacylase is used for resolution of racemic mixtures of amino acids and glucose isomerase, for converting glucose to fructose.

The advantages of immobilized enzymes over soluble enzymes are as follows; i) enzymes can be re-used, ii) continuous process become practical, iii) improved stability of the enzymes, iv) greater purity of the product, v) more favourable or broader environmental stability vi) minimizes effluent problem and material handling. The overall cost of the immobilized enzyme process is only 60% of that of process that uses solubilized enzymes.

At present, the bulk enzyme industry still relies almost entirely on the production of relatively simple enzymes, primarily for application in the food and other industries. By far the majority of these enzymes are hydrolases, such as amylase, cellulase, pectinases, proteases, etc. They are used mostly as additives or processing aids in the baking, dairy, fruit juice and other industries. In contrast to the bulk enzymes, the market for the individual enzymes glucose isomerase, glucose oxidase etc, is expanding rapidly. This fits also for highly purified enzymes used in the pharmaceutical industry, therapeutic application, and clinical and chemical analysis.

Many exciting changes are happening in food industry as a result of improvements and innovations in the use of enzymes either endogenous to food or added exogenously. Endogenous enzyme reactions that are beneficial to food processing such as malting, germination, fermentation are encouraged, while reactions with deleterious effects such as enzymatic browning and off-flavour are controlled. Exogenously added enzymes are used to modify textural, rheological and sensory quality of food products. Certain added enzymes function as antioxidants, preservatives, macerating aid, oil extracting agent, peeling and slicing aid. The preference for natural, safe and eco-friendly products, to which enzymes belongs, widened the scope of use of enzymes in food industry.

QUESTIONS

1. Give an account on microbial enzymes and their applications.
2. Describe the industrial production of amylases and their uses.
3. Explain the enzymatic derivation of fructose from starch.
4. List out various microbial enzymes, organisms producing them and their applications.
5. What are the uses of proteases in food processing industry?
6. What are the uses of lipases?
7. Discuss the applications of immobilized enzymes.
8. Write a note on Use of microbial enzymes in food processing.
9. Which of the following microbial enzymes are used in clarification of fruit juices?

   a) Amylases b) Proteases

   c) Pectinases d) Cellulases
10. Which microbial enzyme is used in improvement of filterability of cane sugar *via* break down of starch in juice

    a) Amylases b) Proteases

    c) Pectinases d) Invertases

9

CONTAMINATION AND SPOILAGE OF FOODS

Microbial contamination sources: Foods may be contaminated by microorganisms at any time during harvest, storage, processing, distribution, handling, or preparation. The primary sources of microbial contamination are soil, air, animal feed, animal hides and intestines, plant surfaces, sewage, and food processing machinery or utensils. Since most foods either carry or eventually acquire bacteria, molds, or yeasts, microorganisms are the major causes of food spoilage. The increased load of microorganisms leads to rapid spoilage.

Organisms: Bacteria and fungi (yeasts and molds) are the principal types of microorganisms that cause food spoilage and food- borne illnesses. The typical microorganisms that cause food spoilage are bacteria (*e.g., Pseudomonas*), yeasts (*e.g., Candida*), and molds (*e.g., Rhizopus*). The growth of microorganisms in the food will increase the load of microorganisms, and the heavier load will add to the difficulty of preventing spoilage and may influence the kind of spoilage to be anticipated.

Food spoilage may be defined as any change that renders food unfit for human consumption. These changes may be caused by various factors, including contamination by microorganisms, infestation by insects, or degradation by endogenous enzymes (those present naturally in the food). In addition, physical and chemical changes, such as the tearing of plant or animal tissues or the oxidation of certain constituents of food, may promote food spoilage. Foods obtained from plant or animal sources begin to spoil soon after harvest or slaughter. The enzymes contained in the cells of plant and animal tissues may be released as a result of any mechanical damage inflicted during post-harvest handling. These enzymes begin to break down the cellular material. The chemical reactions catalyzed by the enzymes result in the degradation of food quality, such as the development of off-flavours, the deterioration of texture, and the loss of nutrients. The enzymatic changes in foods are desirable to some extent as in the case of ripening of fruits, but continued enzymatic activity beyond an optimum point becomes food deterioration due to rotting. The native enzymes may be inactivated by heat, radiation or by the use of specific chemicals.

Classification of food by ease of spoilage

Non-perishable or stable foods: These foods, which do not spoil unless handled carelessly, include such products as sugar, flour, cereals and dry beans.

Semi-perishable foods: If these foods are properly handled and stored, they will remain unspoiled for fairly long period. e.g. potato, onion.

Perishable foods: This group includes most of our important daily foods that spoil readily unless special preservation methods are used. Fruits and vegetables, meat, fish, poultry, eggs and milk.

Chemical changes caused by microorganisms: Food spoilage is caused by the growth of microorganisms, primarily bacteria and fungi that convert nutrients into energy which they use for their own growth. Depletion of nutrient content of food as well as secretion of byproducts from this biochemical processes are two things that contribute to the spoilage of food.

a) **Protein:** Microorganisms produce proteases to hydrolyze the protein in foods to polypeptides, simpler peptides or amino acids before they can serve as nitrogenous food for them. Proteases catalyze the hydrolysis of protein to peptides, which may give a bitter taste to foods. Peptidase catalyze the hydrolysis of polypeptides to simpler peptides and finally to amino acids that give a desirable (cheese) or undesirable flavour.

Anaerobic decomposition of proteins, polypeptides, simpler peptides or amino acids may result in the production of obnoxious odours and is then termed as **putrefaction.** It results in the production of foul smelling sulphur containing products such as hydrogen sulphide, mercaptans, plus ammonia, amines, indole and fatty acids. Products from microbial decomposition of amino acid glycine by different bacterial strains are given below

GLYCINE → Glyoxylic acid + acetic acid + ammonia, by ***E. coli***

GLYCINE → Methylamine + carbon dioxide, by ***Pseudomonas***

GLYCINE → Acetic acid + ammonia + methane, by ***Clostridium***

b) **Carbohydratess** are preferred by microorganisms to other energy yielding foods. Complex polysaccharides are hydrolyzed to simple sugars (glucose) by microorganisms before utilization. Glucose is aerobically oxidized to carbon dioxide and water and anaerobically undergoes decomposition by different types of fermentations. A variety of other products is possible from fermentation of sugars when different microorganisms are active, including fatty acids, organic acids, aldehydes and ketones. The various types of fermentations by different groups of microorganisms and the products of glucose are given below (Table 9.1).

TABLE 9.1 PRODUCTS OF GLUCOSE FERMENTATION

FERMENTATION	PRODUCTS OF GLUCOSE
Alcoholic fermentation by yeast	Ethanol, carbon dioxide
Homofermentative lactic acid bacteria	Lactic acid
Heterofermentative lactic acid bacteria	Lactic, acetic, ethanol, glycerol, CO_2
Coliform type of fermentation	Lactic, acetic, formic, ethanol, CO_2
Propionic fermentation	Propionic, acetic, succinic, CO_2
Butyric -butyl fermentation	butyric, acetic, CO_2, acetone, butanol

c) **Organic acids:** Many of the organic acids usually occurring in foods as salts are oxidized by microorganisms to carbonates, causing the medium to become more alkaline. Aerobically the organic acids may be oxidized completely to carbon dioxide and water. Acids may be oxidized to other simple acids or other products similar to those from sugars. Saturated fatty acids are degraded to acetic acid aided by coenzyme A and unsaturated acids converted to saturated acid for complete beta oxidation.

d) **Alcohols:** Alcohols are oxidized to the corresponding organic acid, e.g. ethanol to acetic acid. Glycerol may be dissimilated to products similar to those from glucose. Acetaldehyde may be oxidized to acetic acid or reduced to ethanol.

Oxidation: $CH_3CH_2OH \rightarrow CH_3CHO \rightarrow CH_3COOH$

e) **Lipids:** Fats are hydrolyzed by microbial lipase to glycerol and fatty acids, which are then dissimilated. Phospholipids may be degraded to their constituent phosphate, glycerol, fatty acids and nitrogenous base.

f) **Pectic substances:** The water insoluble parent pectic substance in plants, protopectin is converted to pectin, a water soluble polymer of galacturonic acid. Pectin esterase causes hydrolysis of the methyl ester linkage of pectin to yield pectic acid and methanol. Polygalacturonases destroy the linkage between galacturonic acid units of pectin to yield smaller chains and ultimately frees galacturonic acid, which may be degraded to simple sugars.

QUESTIONS

1. What are the principal types of microorganisms that cause food spoilage?
2. Give the classification of food by ease of spoilage.
3. Define food spoilage and give examples
4. Discuss the chemical changes caused by microorganisms in various food substrates.
5. What is putrefaction?
6. Fish and eggs represent foods that fall in the category of being

   a) Non-perishable b) Semi-perishable

   c) Perishable d) None of the above
7. Pectinolytic microbes cause the spoilage in

   a) Tubers b) Fruits

   c) Leaves d) Eggs

10

CONTAMINATION, PRESERVATION AND SPOILAGE OF CEREALS AND CEREAL PRODUCTS, SUGAR AND SUGAR PRODUCTS

10.1 CEREALS AND CEREAL PRODUCTS

CONTAMINATION: The sources of contamination in cereals and cereal products are soil, insects, air, etc. Most of the microorganisms are removed with the outer portions of the grain during milling. The bleaching agents used during milling also reduce numbers of microorganisms. The bacteria in wheat flour include *Bacillus, Micrococcus, Flavobacterium.* Mold spores are chiefly of *Aspergillus* spp and *Penicillium* spp and in addition *Alternaria, Cladosporium* and other genera are found. The numbers of microbes in cereals and cereal products vary widely from few hundred per gram to million. Corn meal contains several hundred to several thousand bacteria and mold per gram.

Fig. 10.1 Bread Mold *Mucor*

Bakery products are an important part of a balanced diet and, today, a wide variety of such products can be found on supermarket shelves. This includes unsweetened goods (bread, rolls, buns, crumpets), sweet goods (pancakes, doughnuts, waffles and cookies) and filled goods (fruit and meat pies, sausage rolls, pastries, sandwiches, cream cakes and pizza). However, bakery products, like many processed foods, are subject to physical, chemical and microbiological spoilage. While physical and chemical spoilage limits the shelf life of low and intermediate moisture bakery products, microbiological spoilage by bacteria, yeast and molds is the concern in high moisture products i.e., products with a water activity (aw) > 0.85. Bread is contaminated by mold spores (Fig. 10.1) in air during cooling and wrapping. The contamination of cereals and cereal products with molds have become of significant concern because of mycotoxins such as aflatoxin produced by *Aspergillus flavus*.

PRESERVATION

The low water activity of cereals and cereal products protect them from microbial attack. A storage temperature of 4 to 7°C is recommended for dry products. The bakery products are preserved by asepsis during slicing and packing , use of heat, use of freezing storage, use of chemical preservatives like ammonia 2%, propionic acid 1% reduce mold growth in high moisture products. UV radiation is used to destroy or reduce the number of mold spores in cereals and cereal products.

To obtain maximum shelf life free of mold spoilage, high levels of sanitation must be maintained in baking and packing areas. Oven heat destroys all fungal life-forms, and any spoilage by these organisms is due to re-inoculation after baking. A number of compounds have been proposed for use as fungistats in bread. Some have proved to be innocuous to molds, toxic to humans, or both. Soluble salts of propionic acid, principally sodium propionate, have been accepted and extend shelf life appreciably in the absence of a massive inoculum. Sorbic acid or potassium sorbate and acetic acid also have a protective effect.

Bakery products are subject to the microbiological spoilage problems affecting other foods. If moisture content is kept below 12 to 14 percent (depending on the composition), growth of yeast, bacteria, and molds is completely inhibited. Nearly all crackers and cookies fall below this level, although jams, marshmallow, and other adjuncts may be far higher in moisture content. Breads, cakes, sweet rolls, and some other bakery foods may contain as much as 38 to 40 per cent water when freshly baked and are subject to attack by many fungi and a few bacteria (Table 10.1).

TABLE 10.1 SPOILAGE OF CEREALS AND CEREAL PRODUCTS

PRODUCT	SPOILAGE	ORGANISM
CEREAL grains and Meals	Acid Fermentation	Lactic acid bacteria, Coliform bacteria
	Alcoholic Fermentation	Yeasts
	Mold growth	*Aspergillus, Penicillium, Fusarium*
BREAD	Moldy growth	Bread mold- *Rhizopus, Mucor, Aspergillus niger, Penicillium expansum*
	Ropiness	*Bacillus subtilis*
	Red bread	Pigmented bacteria-*Serratia marcescens*

The only widespread food poisoning in which bread has been a vector has resulted from ergot, a fungus infection of the rye plant caused by *Claviceps purpurea*. Ergot contamination of bread made from rye, or from blends of rye and wheat, has caused epidemics leading to numerous deaths.

10.2. SUGAR AND SUGAR PRODUCTS

CONTAMINATION

Sugar is contaminated with microorganisms from sugarcane and soil. The contaminants are *Leuconostoc, Bacillus, Micrococcus, Flavobacterium, Alcaligenes, Enterobacter* and yeasts. In raw sugar 400 to 68,000 microorganisms per gram have been found. The chief sources of contamination in honey are the nectar of flower and the honey bee, yeasts come from these sources. Candies from retail market contain 0 to 2 million bacteria per piece. Candies receive most of their contamination from their ingredients and added by air, dust and handling. Candies were associated with the cases of salmonellosis.

PRESERVATION

Manufacturing processes such as clarification, evaporation, crystallization, centrifugation and filtration reduce the number of microorganisms present in sugar and sugar products. Boiling process during evaporation kills the spoilage microorganisms. Chemical preservatives are effective in reducing the number of microorganisms. Microorganisms associated with spoilage of sugar and sugar products are listed in Table 10.2.

TABLE 10.2. SPOILAGE OF SUGARS AND SUGAR PRODUCTS

Product	Spoilage	Organism
Sucrose- raw juice	Slime formation	*Leuconostoc mesenteroides*, *Leuconostoc dextranicum*
Liquid sugar	Yeast growth	*Saccharomyces, Candida, Rhodotorula*
Syrup	Ropy sap	*Enterobacter aerogenes*
	Red sap	*Micrococcus roseus*
	Cloudy sap	*Pseudomonas fluorescens*
Molasses	Gas production	Osmophilic yeasts
Honey		*Zygosaccharomyces*
Candy		*Clostridium* sp
Chocolate	Explosion	*Clostridium sporogenes*

QUESTIONS

1. Discuss the contamination, preservation and spoilage of cereals and cereal products.
2. Discuss the contamination, preservation and spoilage of sugar and sugar products.
3. Write short notes on
   i) Chemical preservation of cereals products
   ii) Ropiness in bread
   iii) Chocolate explosion.

11

CONTAMINATION, PRESERVATION AND SPOILAGE OF FRUITS AND VEGETABLES

11.1. FRUITS-Maturation, Ripening and Storage

Fruits are natural foods containing substantial quantities of essential nutrients and the best sources of minerals like sodium, potassium and magnesium, vitamins (antioxidants), such as vitamin A, C and enzymes. They are rich in fructose and water. Fruits are living biological entities that perform a number of metabolic functions. Two functions of particular importance in fruit processing are respiration (the breaking down of carbohydrates, giving off carbon dioxide and heat) and transpiration (the giving off of moisture). Senescence is the period when chemical synthesizing pathways give way to degradative processes, leading to aging and death of tissue. Fruit ripening is thus the result of many complex changes, some interactive but many independent of one another. Once the fruit is harvested, respiration and transpiration continue, but only for as long as the fruit can draw on its own food reserves and moisture. It is this limited ability to continue vital metabolic functions that defines fruit as perishable.

Storage concerns

As harvested fruit ages, it is particularly important to manage the temperatures under which it is stored. For example, respiration largely involves enzymatic processes, which are significantly controlled by ambient temperature. The rate of chemical change in fruit generally doubles for every increase of 10°C (at room temperature).

CONTAMINATION

Microorganisms can also cause problems during senescence and storage. Many bacteria and fungi, for instance, are involved in decay after harvest. Typical fungi include *Alternaria, Botrytis, Monilia,*

Penicillium, and *Rhizopus*. These fungi are generally weak pathogens, in that they usually infest only weak or damaged fruit. Efforts to control infection begin in the orchard, usually with the application of fungicides. Cooling of the fruit or, conversely, hot-water dipping may also enhance storage quality. In addition, the careful application of ionizing radiation has been shown to inhibit microbial growth.

PRESERVATION OF FRUITS

Since fruits are generally acidic, they are naturally amenable to preservation. The premier role of acidity in preservation is to stop bacterial growth. Second, increased acidity can activate chemical reactions such as pectin set, which lowers water activity and reduces the possibility of microbial growth.

1. **ASEPSIS:** Removal of microorganisms. Fruit is usually washed prior to any processing. Washing is typically conducted with a high-pressure soak or spray system. Under some conditions a surfactant or detergent may be added in order to release stubborn soil attached to the fruit. In apple processing a high-quality wash is necessary to ensure the safe removal of microorganisms responsible for mycotoxin formation and possible gastrointestinal poisoning.

2. **PACKAGING:** Packaging systems for fresh fruit usually involve a simple plastic breathable bag or over wrap. However, as the market value of high-quality fruit has increased, so too have efforts to develop improved packaging. These efforts have been primarily in the area of modified-atmosphere packaging (MAP). In this type of packaging the barrier properties of the material are carefully selected according to the respiration characteristics of the fruit. The goal is to allow an exchange of gases and moisture that produces the optimal storage environment. Continued work in this field is producing "smart" films, which not only produce the optimal atmosphere for storage but also change their barrier properties depending on the ambient temperature and on the respiration rate of the fruit.

3. **DEHYDRATION:** Dehydration is among the oldest and most common forms of fruit preservation. In dehydration, moisture in the fruit is driven off; leaving a stable food that has moisture content below that at which microorganisms can grow. There are three basic systems for dehydration: sun drying, such as that used for raisins; hot-air dehydration; and freeze-drying.

Dehydration has a number of advantages. Dehydrated fruit has a virtually unlimited shelf life when held under proper storage conditions. Drying does not significantly reduce the calories or minerals, and vitamin losses are similar to other preservation methods. In addition, by reducing the weight and the need for refrigeration, handling and transportation costs can be reduced dramatically. Dehydrated fruits are typically reduced in weight by 75 to 90 percent. Although dehydration offers a convenient product form, it usually requires a careful inactivation of enzymes. This is usually accomplished by blanching of the fruit or by chemical inactivation. Typically, sulfur dioxide is added for its antioxidant and preservative effects. In order to control browning, the fruit is often treated prior to dehydration with sodium sulfite and sodium bisulfite.

4. **CANNING:** In thermal processing, heat is used to destroy spoilage organisms and to inactivate troublesome enzymes. Enzymes are typically responsible for browning, softening, and the development of off-flavours. For high-acid fruit products the most typical thermal process is canning, in which fruit or fruit products are hot-filled or heated in a hermetically sealed container. The process temperature is generally in the range of 88° C .In general the more acid the fruit, the less heat required for its preservation.

5. **LOW TEMPERATURE:** Once harvested, fruits are moved to storage. In the case of highly heat-sensitive products such as raspberries or cherries, the fruit should be pre-cooled prior to storage. Pre-cooling can be accomplished by hydro-cooling (immersion of the fruit in cold water) or vacuum cooling (moistening and then placing under vacuum in order to induce evaporative cooling). Apples and grape fruits store well at 10°C and 11°C respectively.

A typical storage system for fruit is cold storage, using refrigerated air. Other techniques include **controlled-atmosphere** (CA) storage and hypobaric storage. In CA storage the oxygen and carbon dioxide content of the storage environment are controlled in such a way as to retard senescence and further deterioration of the fruit. In general, oxygen levels are reduced and carbon dioxide levels increased. CA conditions can be generated in a number of ways. Conventional CA depends on the respiration of the fruit to generate carbon dioxide, and the concentration of this gas is controlled by wet scrubbers, hydrated lime, or other commercial carbon dioxide removal systems. Liquid nitrogen and compressed nitrogen gas have also been used to flush out the ambient air of the storage facility.

In other systems, oxygen is converted to carbon dioxide by reaction with liquid propane or by catalytic burning. Hypobaric storage involves the cold storage of fruit under partial vacuum. Typical conditions include pressures as low as 80 and 40 millimetres of mercury and temperatures of 5°C . Hypobaric conditions reduce ethylene production and respiration rates; the result is an extraordinarily high-quality fruit even after months of storage.

6. **FREEZING:** Freezing of fruits and fruit products is a common consumer practice. Cold temperatures act to retard the spoilage of fruit by inhibiting microbial action and slowing metabolic processes. In order to achieve extended storage life, the product must be held well below the freezing point of water-typically at a minimum of -23°C. Generally, rapid freezing leads to an improved texture upon thawing. A prerequisite for effective freezing is inactivation of fruit enzymes. Traditionally, this is done through blanching or by the addition of a chemical. Blanching consists of heating the fruit for a short time in water or steam prior to cooling and subsequent freezing. The blanch step is intended to inactivate enzyme systems responsible for off-flavours, browning, and softening.

Fruit juice: After fresh fruit, one of the most common fruit products is fruit juice. Fruit juice can take on many forms, including a natural-style cloudy product, "nectar"-type product containing suspended solids, a fully clarified juice, juice concentrate, and fruit drinks. The processing of fruit juice involves washing, extraction, clarification, and preservation.

7. **CHEMICAL PRESERVATION:** Chemicals also can be used as a preservative, either through artificial addition or through the action of microorganisms. An example of the latter method is yeast fermentation, which can cause an increase in ethyl alcohol sufficient to preserve the fruit product. Pickling is another example of chemical preservation. In the case of pickling, the product may be preserved by the addition of salt, sugar, acetic acid (vinegar), and alcohol. High sugar content also acts as a fruit preservative by tying up all available moisture so that microorganisms cannot grow.

SPOILAGE OF FRUITS: The post-harvest infection of fruits by microorganisms may cause serious losses of fruits both in terms of quantity and quality. Initial infection which may be the direct result of mechanical or insect damage is followed by invasion of fruit tissues by secondary infection by common saprophytic organisms. Many bacteria and fungi, for instance, are involved in decay after harvest (Table 11.1). Typical fungi include *Alternaria, Botrytis, Colletotrichum, Fusarium, Monilia, Penicillium,* and *Rhizopus*. These fungi are generally weak pathogens, in that they usually infest only weak or damaged fruit. Molds are the most predominant spoilage organisms in fruits. Fungal rots in fruits like apple show brown or cream coloured areas (Fig 11.1). Bacterial soft rot caused by *Erwinia carotovora* is wide spread in fruits that are not very acid. Fruits like straw berries, melons are in direct contact with moist soil and become infected from that source.

Fig 11.1 Fruit-Apple spoilage

TABLE 11.1. MICROBIAL SPOILAGE OF FRUITS

FRUIT	SPOILAGE	ORGANISM
Mango	Black mold rot	*Aspergillus niger*
	Anthracnose	*Colletortichum* sp.
Pine apple	Chocolate rot	*Sclerotium rolfeit*
	Soft rot	Yeast injury
Pappaya	Soft rot	*Botryodiplodia theobromae*
	Anthracnose	*Colletotrichum gloeosporoides*
Banana	Anthracnose	*Colletotrichum musae*
	Black mold rot	*Aspergillus niger*
Apple	Black rot	*Botryosphaeria btuse*
	Circular brown rot	*Alternaria alternata*
	Sour rot	*Geotrichum candidum*
	Bacterial soft rot	*Erwinia caratovora*
Mandrian orange	Dirty core rot	*Absidia corymbifera*
	Dark brown end rot	*Alternaria citri*
	Mold rot	*Penicillium crustosum*
Sapota	Black mold rot	*Aspergillus niger*
Grapes	Brown discolouration	*Alternaria alternata*
	Green ball rot	*Cladosporium* sp.
	Blue mold rot	*Penicillium citranum, P. frequentans*

11.2. VEGETABLES

CONTAMINATION

Vegetables are perishable products with active metabolism during post-harvest period. Therefore proper handling and environmental conditions after harvesting are essential in maintaining product quality and availability. The quality of vegetables seldom improves during the post-harvest period. Vegetables, due to their neutral pH are susceptible to both fungi and bacteria. Several intrinsic factors predispose fresh vegetables to microbial attack and subsequent spoilage. The high water content, nutrient composition and pH of most vegetables make them capable of supporting the growth of many bacteria, yeasts and molds.

The location of plant part used as vegetable is an important factor in microbial contamination and spoilage. The underground parts like roots, tubers or bulbs as in radish, beet, carrot, potato, onion are in direct contact with moist soil and become form the source. Vegetables like cucumbers, tomatoes may be in direct contact with the surface soil. Leaves, stems and flowers as in lettuce, greens, cabbage, asparagus, cauliflower are generally infected by plant pathogens from the field. The close proximity of the plant to the soil as well as presence of natural openings (lenticels and stomata), through which the microorganisms can penetrate, allow for easy contamination of vegetables. Wounds resulting from harvesting and handling such as stem scars, scratches, abrasions and bruises also provide ports of entry for invading microorganisms.

PRESERVATION

1. **ASEPSIS:** Boxes, lugs, baskets and other containers used between harvesting and processing should be practically free from the growth of microorganisms. Contamination from process equipment can be reduced by cleaning and sanitizing. Washing of vegetables with wash water of good bacteriological quality removes casual contamination organisms.

2. **PICKLING:** In both fermented and pickled vegetables, acid is used to preserve the products. Pickled vegetables include cucumbers, green tomatoes, onions, radishes, and cabbages. The amount of salt varies from 2.25 to 2.5% in making sauerkraut. The variety of vegetables used for fermentation or pickling may not be the same as fresh market vegetables. Owing to the acid environment, fermented or pickled vegetables need less heat treatment before

being placed in containers. Vegetables high in protein - green peas, lima bean and that soften readily - onion, cauliflower are preserved by addition of 18.6 to 21.2% sodium chloride.

3. **CANNING:** Putting foods into metal cans or glass jars is the major food-processing method of the world. It is particularly useful in developing countries where refrigeration is limited or nonexistent. In the canning process, vegetables are often cut into pieces, packed in cans, and put through severe heat treatment to ensure the destruction of bacterial spores. The containers are sealed while hot so as to create a vacuum inside when they are cooled to room temperature. Properly processed canned vegetables can be stored at room temperature for years. Minor defects of the process, however, will result in bulged cans after long periods of storage. For safety reasons, the contents of these cans should not be consumed. Although in most cases bulged cans are caused by the formation of gas from chemical reactions between the metal cans and their acidic contents, there is a remote possibility that inadequate heat processing did not destroy all bacterial spores. And, eventhough most heat-resistant spores are nonpathogenic, spores of *Clostridium botulinum* can survive under-processing and produce deadly toxins that cause botulism.

Unfortunately, because of the severe heat treatment, some canned vegetables can have inferior quality and less nutritive value than fresh and frozen products. The nutrient most susceptible to destruction in canning is vitamin C. For high-quality products, aseptic canning is practiced. Also known as high-temperature–short-time (HTST) processing, aseptic canning is a process whereby pre-sterilized containers are filled with a sterilized and cooled product and sealed in a sterile atmosphere with a sterile cover. The process avoids the slow heat penetration inherent in the traditional tin- container heating process, thus creating products of superior quality.

The canning process can be illustrated by the example of green beans (*Phaseolus vulgaris* L.). After arrival at the processing plant, the beans are conveyed to size graders. Graders consist of revolving cylinders with slots of various diameters through which the beans fall onto conveyers. The conveyers carry them to snipping machines, where their tips and stems are cut off. The snipped beans then pass over inspection belts, where defective beans are removed. Smaller beans are canned as whole beans, while larger beans are cut crosswise by machine into various lengths. Some smaller beans are cut lengthwise and marketed as French-cut beans. Both the small whole cut beans are blanched for 1.5 to 2 minutes in 82°C water and mechanically packed in cans. The cans are then filled with hot water and dry salt or with brine, steam-exhausted for approximately five minutes, and sealed while hot or with steam flow. Depending on the size of the can, they are heat-processed for various periods of time-from 12 minutes at 120°C to 36 minutes at 115°C. The cans are cooled to room temperature, labeled, and packaged for storage or immediate distribution.

4. **IRRADIATION:** Ionizing radiation, mostly gamma-ray, has been used in several countries to preserve vegetables. The practice is quite common in preventing potatoes from sprouting during long-term storage. Despite studies showing that products treated with low-dose ionizing radiation are safe, consumers are still concerned about this processing technology and have not accepted it.
5. **COOL STORAGE:** Bulk storage of fruits and vegetables in cool environment created by air cooler or water spray is widely practiced in major fruits and vegetable growing areas. Banana and tomato are stored at 13°C to slow down natural ripening. Potato and cabbage are stored at 15°C. Loss of crispness of fruits and vegetables, loss of flavour, loss of nutrients-sugars in sweet corn are few problems associated with cool storage.
6. **FREEZING:** Frozen foods have outstanding quality and nutritive value. Indeed, some frozen vegetables, such as green peas and sweet corn, may be superior in flavour to fresh produce. The high quality of frozen foods is mainly due to the development of a technology known as the individually quick-frozen (IQF) method. IQF is a method that does not allow large ice crystals to form in vegetable cells. Also, since each piece is individually frozen, particles do not cohere, and the final product is not frozen into a solid block. Various freezing techniques are commonly used in the preservation of vegetables. These include blast freezing, plate freezing, belt-tunnel freezing, fluidized-bed freezing, cryogenic freezing, and dehydro-freezing. The choice of method depends on the quality of end product desired, the kind of vegetable to be frozen, capital limitations, and whether or not the products are to be stored as bulk or as individual retail packages.

Most vegetables frozen commercially are intended for direct consumer use or for further processing into soups, prepared meals, or specialty items. Advances in packaging materials and techniques have led to bulk frozen products being stored in large retortable pouches. Many restaurants and institutions prefer bulk frozen soups packaged in these pouches because of their quality and convenience.

One of the most important vegetable crops preserved by freezing is sweet corn (*Zea mays* L.). Both corns on the cob and cut corn are frozen. Sweet corn must be harvested while still young and tender and while the kernels are full of "milk." After the ears are mechanically harvested, they are promptly hauled to the processing plant, where they are automatically dehusked and desilked. Probably more than any other vegetable, sweet corn loses its quality rapidly after harvest. Frozen corn maintains high quality by being processed within a few hours of picking. Corn on the cob is a particularly difficult vegetable to freeze. The dehusked and desilked ears are thoroughly washed and blanched in steam for 6 to 11 minutes and then promptly cooled. However, even an 11-minute blanch in steam does not completely inactivate all the enzymes in the cob portion. It is believed that the off-flavour frequently found in home-frozen corn on the cob comes from off-flavours produced in the cob that migrates out to the kernels. Blanched and cooled corn is quickly frozen by the fluidized- bed freezing process before packing. Blanched whole-kernel corn is produced either by blanching the corn on the cob before cutting; by partially blanching on the cob to set the milk, then cutting and blanching again; or by cutting before blanching. The "split" method of blanching twice produces the highest-quality product. After the corn is cut, impurities such as husk, silk, and imperfect kernels must be removed by either brine flotation or froth washing. In both methods the sound corn stays at the bottom while the impurities float off the tank. Whole-kernel corn can be frozen quickly using the individually quick-frozen method. Frozen corn can be packaged into polyethylene bags or cardboard cartons and labeled for retail, or it can be bulk-stored for further processing into components of value-added products such as frozen dinners.

TABLE 11.2. MAJOR VEGETABLES AND KINDS OF PROCESSING

	CANNING	FREEZING	DEHYDRATION	PICKING
Bean	+	+	+	-
Cabbage	-	-	+	+
Carrot	+	+	+	+
Cucumber	-	-	-	+
Lima bean	+	+	-	+
Onion	-	-	+	+
Pea	+	+	-	-
Potato	-	+	+	-
Spinach	+	+	-	-
Sweet corn	+	+	-	-
Sweet potato	+	-	+	-
Tomato	+	-	+	-

SPOILAGE OF VEGETABLES: Vegetables are exposed to microbial spoilage (Table 11.3) depending on several factors such as cell wall thickness, moisture content, inherent chemical substances etc. Vegetables are moisture rich food, relatively more prone to microbial attack. The location of plant part used as vegetable is an important factor in microbial contamination and spoilage. The underground parts like roots, tubers or bulbs as in radish, beet, carrot, potato, onion are in direct contact with moist soil and become form the source. Vegetables like cucumbers, tomatoes (Fig. 11.2) may be in direct contact with the surface soil. Leaves, stems and flowers as in lettuce, greens, cabbage, asparagus, and cauliflower are generally infected by plant pathogens from the field.

Fig. 11.2 Vegetable-tomato spoilage

TABLE 11.3. MICROBIAL SPOILAGE OF VEGETABLES

SPOILAGE	ORGANISM	VEGETABLES
Sour rot/Watery soft rot	*Geotrichum candidum*	Bean, carrot, cabbage, onion, tomato, cucumber
Rhizopus soft rot	*Rhizopus stolonifier*	Carrot, potato, cabbage, radish
Black rot	*Xanthomonas campestris*	Cabbage, cauliflower
Bacterial soft rot	*Erwinia carotovora*	Carrot, cabbage, green beans
Gray mold rot	*Botrytis cinerae*	Carrot, cabbage, green beans
Black rot	*Alternaria* sp.	Carrot, cabbage, green beans
Blue mold rot	*Penicillium* sp.	Onions
Anthracnose	*Colletotrichum* sp.	Green beans, tomato, onion
Fusarium rot	*Fusarium* sp.	Carrot, potato

QUESTIONS

1. Discuss in detail the various methods of preservation of fruits.
2. Write a brief account on controlled atmosphere and modified atmosphere in storage of fruits.
3. Describe freezing of fruits and fruit products.
4. List out various types of spoilage of fruits and the causative organisms.

5. What are the different methods of preservation of vegetables?
6. Write a brief account on contamination of vegetables
7. Write on cool storage and freezing of vegetables
8. Explain major vegetables and the types of processing.
9. Write an account of microbial spoilage of vegetables.
10. Write short notes on
    i) Hydrocooling
    ii) Dehydration of fruits
    iii) Pickling of vegetables
    iv) Canning of green beans
    v) Freezing of sweet corn.
    vi) Gray mold rot
    vii) Bacterial soft rot.

12

CONTAMINATION, PRESERVATION AND SPOILAGE OF MEAT AND MEAT PRODUCTS

CONTAMINATION

The healthy inner flesh of meat contains few or no microorganisms, although they have been found in lymph nodes, bone marrow and intestine. *Staphylococcus, Streptococcus, Clostridium* and *Salmonella* have been isolated from lymph nodes of red meat animals. During slaughtering the lymph nodes are removed from edible parts. The important contamination, however comes from external sources during bleeding, handling and processing. During bleeding, skinning and cutting the main sources of microorganisms are from the exterior of the animal- hide, hoof and hair and the intestinal tract. The exterior of the animal harbours large number and many kinds of microorganisms from soil, water, feed and manure as well as its natural surface flora.

Knives, cloths, air and hands and clothing of workers can serve as intermediate sources of contamination. During handling and processing the contamination can come from containers, air, other contaminated meat, equipments and personnel.

Molds that grow on the surface of the meat include the species of *Cladosporium, Sporotrichum, Geotrichum, Mucor, Monilia, Penicillium* and *Alternaria.* The important bacteria found in meat are *Staphylococcus, Sarcina, Streptococcus, Clostridium, Leuconostoc, Lactobacillus, Proteus, Flavobacterium, Bacillus, Streptomyces, Escherichia* and *Salmonella.*

PRESERVATION

The preservation of meats, as of most perishable foods, usually is accomplished by a combination of preservative methods. The preservation of meat is more difficult than most kinds of foods because, it is a good culture medium for microbial growth and contains large number of spoilage microorganisms.

Preservation and storage: Meat preservation helps to control spoilage by inhibiting the growth of microorganisms, slowing enzymatic activity, and preventing the oxidation of fatty acids that promote rancidity. There are many factors affecting the length of time meat products can be stored while maintaining product safety and quality. The physical state of meat plays a role in the number of microorganisms that can grow on meat. For example, grinding meat increases the surface area, releases moisture and nutrients from the muscle fibres, and distributes surface microorganisms throughout the meat. Chemical properties of meat, such as pH and moisture content, affect the ability of microorganisms to grow on meat. Natural protective tissues (fat or skin) can prevent microbial contamination, dehydration, or other detrimental changes. Covering meats with paper or protective plastic films prevents excessive moisture loss and microbial contamination.

1. **Cold storage:** Temperature is the most important factor influencing bacterial growth. Pathogenic bacteria do not grow well in temperatures under 3°C. Therefore, meat should be stored at temperatures that are as cold as possible. Refrigerated storage is the most common method of meat preservation. The typical refrigerated storage life for fresh meats is 5 to 7 days. Freezer storage is an excellent method of meat preservation. It is important to wrap frozen meats closely in packaging that limits air contact with the meat in order to prevent moisture loss during storage. The length of time meats are held at frozen storage also determines product quality. Under typical freezer storage of -18°C beef can be stored for 6 to 12 months, lamb for 6 to 9 months, pork for 6 months, and sausage products for 2 months.

2. **Freezing:** The rate of freezing is very important in maintaining meat quality. Rapid freezing is superior; if meats are frozen slowly, large ice crystals form in the meat and rupture cell membranes. When this meat is thawed, much of the original moisture found in the meat is lost as purge (juices that flow from the meat). For this reason cryogenic freezing (the use of supercold substances such as liquid nitrogen) or other rapid methods of freezing meats are used at the commercial level to maintain maximal product quality. It is important to note, however, that freezing does not kill most microorganisms; they simply become dormant. When the meat is thawed, the spoilage continues where it left off.

Thawing meats often can cause more detrimental quality changes than freezing. In contrast to freezing, thawing should be a slow process. Meats are best thawed in the refrigerator with packaging left intact, so that moisture loss is minimized. Placing frozen meats out on a warm countertop or under warm water subjects the meat's outer layers to room temperatures for long periods of time before the meat is ready for cooking (completely thawed). This rapid method provides a suitable environment for the growth of food-borne microorganisms and increases the risk of food poisoning.

3. **Vacuum packaging:** Oxygen is required for many bacteria to grow. For this reason most meats are vacuum-packaged, which extends the storage life under refrigerated conditions to approximately 100 days. In addition, vacuum packaging minimizes the oxidation of unsaturated fatty acids and slows the development of rancid meat.
4. **Canning:** The second most common method of meat preservation is canning. It involves sealing meat in a container and then heating it to destroy all microbes capable of food spoilage. Under normal conditions canned products can safely be stored at room temperature indefinitely. However, certain quality concerns can compel processors or vendors to recommend an optimal "sell by" date.
5. **Drying:** Drying is another common method of meat preservation. Drying removes moisture from meat products so that microorganisms cannot grow. Dry sausages, freeze-dried meats, and jerky products are all examples of dried meats capable of being stored at room temperature without rapid spoilage.
6. **Fermentation:** One ancient form of food preservation used in the meat industry is fermentation. Fermentation involves the addition of certain harmless bacteria to meat. These fermenting bacteria produce acid as they grow, lowering the pH of the meat and inhibiting the growth of many pathogenic microorganisms.
7. **Irradiation:** Irradiation, or radurization, is a pasteurization method accomplished by exposing meat to doses of radiation. Radurization is as effective as heat pasteurization in killing food-spoilage microorganisms. Irradiation of meat is accomplished by exposing meat to high-energy ionizing radiation produced either by electron accelerators or by exposure to gamma radiation emitting substances such as cobalt-60 or cesium-137. Irradiated products are virtually identical in character to non-irradiated products, but they have significantly lower microbial contamination. Irradiated fresh meat products still require refrigeration and packaging to prevent spoilage, but the refrigerated storage life of these products is greatly extended.
8. **Smoking:** Use of wood smoke as a preservative in meat has two purposes: i) to add desired flavour, ii) to aid in preservation. The

preservatives added to meat, with the action of heat produced during smoking, has a germicidal effect.

9. **Antibiotics:** Antibiotics such as chlortetracycline, oxytetracycline and chloramphenicol are used in meat to prolong the storage life at chilling or higher temperatures. The antibiotics may be applied to meat in various ways
   i) It is fed to the animal over a long period
   ii) It may be fed more intensely for a short period before slaughtering
   iii) It may be infused in to the carcass
   iv) It may be applied to the surface of pieces of meat.

MICROBIAL SPOLAGE OF MEAT: The predominant bacteria associated with spoilage of meat are *Pseudomonas* spp., *Carnobacterium* spp., *Enterobacteriaceae, Lactobacillus* spp., and *Leuconostoc* spp. The main defects in meat are off-odours and off-flavours, but discolouration and gas production also occur. Bacteria associated with the spoilage of refrigerated meat products, causing defects such as sour, off-flavours, discolouration, gas production, slime production and decrease in pH, consist of *B. thermosphacta, Carnobacterium* spp, *Lactobacillus* spp, *Leuconostoc* spp. and *Weissella* spp (Table 12.1).

TABLE 12.1. MICROBIAL SPOILAGE OF MEAT

AEROBIC SPOILAGE	ORGANISMS
1. SURFACE SLIME	*Pseudomonas, Streptococcus, Bacillus, Leuconostoc, Lactobacillus, Proteus, Alcaligenes, Micrococcus.*
2. CHANGE IN COLOUR - Red colour of meat changed to Green, brown or gray	*Leuconostoc, Lactobacillus*
3. CHANGE IN FAT Oxidative rancidity.	Lipolytic bacteria– *Pseudomonas, Achromobacter.*
4. PHOSPHORESCENCE-surface of meat	Luminous bacteria- *Photobacterium* sp
5. SURFACE COLOUR CHANGE	
Due to pigmented bacteria -Red spot	*Serratia marcescens*
Blue colour	*Pseudomonas syncyanea*
Yellow discolour	*Micrococcus. Flavobacterium*
Greenish blue	*Chromobacterium*
Stamping ink disease-purple	*Micrococcus.*
6. OFF FLAVOURS/ OFF TASTES	Acid and gas producing bacteria

7. WHISKERS White fuzzy growth	Mold- *Thamnidium, Mucor, Rhizopus*
8.COLOUR CHANGE due to mold	
BLACK SPOT	*Cladosporum herbarum*
WHITE SPOT	*Sporotrichum carnis*
GREEN PATCHES	*Penicillium expansum, P. oxalicum*
ANAEROBIC SPOILAGE	**ORGANISMS**
1. SOURING caused by organic acids	*Clostridium butyricum*, Coliforms
2. PUTREFACTION- Anaerobic decomposition of Protein with the production of Foul smelling compounds	*Pseudomonas putida, Pseudomonas putrefaciens Clostridium, Alcaligenes, Proteus*

QUESTIONS

1. What the various sources of microbial contamination in meat and meat products?
2. What are the microorganisms found on the surface of meat?
3. Write a detailed account on different methods of preservation of meat.
4. Write on freezing of meats and its shelf life.
5. Describe the use of antibiotics in the preservation of meat.
6. List out various types of bacterial spoilage of meat and meat products
7. List out various types of fungal spoilage of meat and meat products
8. List out various types of aerobic spoilage of meat and meat products
9. List out various types of anaerobic spoilage of meat and meat products
10. Write short notes on - Smoking, Freezer storage of meat, Whiskers, Putrefaction of meat, Phosphorescence, Pigmented bacteria in meat.
11. The anaerobic decomposition of protein in meat is called

    a) Proteolysis b) Putrefaction

    c) Nitrification d) Lipolysis
12. Whiskers in meat are caused by

    a) Bacteria b) Fungi

    c) Viruses d) Protozoa

13

CONTAMINATION, PRESERVATION AND SPOILAGE OF FISH AND SEA FOODS

13.1. CONTAMINATION

MICROFLORA: The flora of living fish depends upon the microbial contents of the waters in which they live. The slime in the outer surface of the fish contain bacteria of the genera: *Pseudomonas, Alcaligenes, Micrococcus, Flavobacterium, Corynebacetrium, Bacillus, Sarcina, Serratia* and *Vibrio*. The same species of bacteria are found in shrimps, crabs, lobsters and similar sea foods. Oysters and Shell fish that pass large amount of water through their bodies, pick-up soil and water microorganisms in this way including human pathogens. Common spoilage microorganisms of fish include species of *Pseudomonas, Moraxella*, and *Acinetobacter*, found mainly in marine fish, and *Bacillus* and *Micrococcus*, found in freshwater fish. Fish may also contain pathogenic (disease-causing) microorganisms such as *Salmonella* and *Escherichia coli*. Pathogenic contamination is of special concern with mollusks because they are often eaten raw and as whole animals.

SOURCES: Boats, boxes, bins, fish houses and fishermen are the sources of contamination of these bacteria. The bacterial population ranges from few hundred to few million per square centimeter. Because of their soft tissues and aquatic environment, fish are extremely susceptible to microbial contamination. At the time of harvest, fish carry a high microbial load on the surface of their skin, in their intestinal tract, and in their gills. The type and number of microorganisms that live in fish vary according to the season, the species, and the natural habitat. Additional contamination may occur during the harvesting, handling, or processing of the fish. Harvested fish must be immediately stored in a low-temperature environment such as ice or refrigerated seawater. This chilling process slows the growth of microorganisms that live in fish and inhibits the activity of enzymes. Because fish have a lower body temperature, softer texture, and less connective tissue than land animals, they are much more susceptible to microbial contamination and structural degradation. If immediate chilling is not possible, then the fish must generally be sold and eaten on the day of the harvest.

13.2. PRESERVATION

Handling of harvested fish: The retention of nutritional properties and product quality of fish is dependent on proper handling of the catch after it has been harvested from its aquatic environment.

PREPROCESSING: Preprocessing of fish prepares the raw material for final processing. It is often performed on shipboard or in a shore-based plant and includes such operations as inspection, washing, sorting, grading, and butchering of the harvested fish. The butchering of fish involves the removal of non-edible portions such as the viscera, head, tail, and fins. Depending on the butchering process, as much as 30 to 70 percent of the fish may be discarded as waste or reduced to cheap animal feed. The lower figure applies when the fish is canned or sold as "whole." The higher figure applies when the fish is filleted or made into other pure meat products; in these cases the skeleton is discarded with as much as 50 percent of the edible flesh attached. Efforts to utilize this discarded fraction for the production of alternative food products have begun in the fish industry.

FINAL PROCESSING: The four basic procedures used in the final processing of fish products are heating, freezing, controlling water activity (by drying or adding chemicals), and irradiating. All these procedures increase the shelf life of the fish by inhibiting the mechanisms that promote spoilage and degradation. Each of these procedures also has an effect on the nutritional properties of the final product.

Chilling: Ice cooling and holding normally requires a one-to- one or one-to-two weight ratio of ice to fish, depending on the specific geographic location and the time it takes to transport the fish to the processing plant. Refrigerated sea water cooling and holding cause less bruising and other structural damage to the fish carcasses than ice cooling. However, fish cooled in refrigerated seawater absorbs salt from the water. For this reason fish that is destined for sale on the fresh or frozen market may be held in refrigerated seawater for only a limited amount of time. The addition of salt during canning or smoking processes is adjusted in order to compensate for any absorbed salt.

Freezing: Of the many processing methods used to preserve fish, only freezing can maintain the flavour and quality of fresh fish. Freezing greatly reduces or halts the biochemical reactions in fish flesh. For instance, in the absence of free water, enzymes cannot react to soften and degrade the flesh. The three steps for freezing fish include immediate cooling and holding, rapid freezing, and cold storage. If fish is frozen improperly, structural integrity may be compromised because of enzymatic degradation, texture changes, and dehydration.

Immediate cooling: The rapid cooling and holding of fish at temperatures between 2 and-2°C takes place immediately after the fish have been harvested.

Rapid freezing: The key to freezing is rapid reduction of the temperature to between -2°C and -7°C. This temperature range represents the zone of maximum ice crystal formation in the cells of the flesh. If water in the cells freezes quickly, then the ice crystals will remain small and cause minimal damage to the cells. However, slow freezing results in the formation of large ice crystals and the rupturing of the cell membranes. When slow- frozen flesh is thawed, the ruptured cells release water (called drip) and many compounds that provide certain flavour characteristics of fish, resulting in a dry, tasteless product. Fish that passes through the zone of maximum ice crystal formation in less than one hour will generally have minimum drip loss upon thawing.

Cold storage: Once fish is frozen, it must be stored at a constant temperature of -23°C or below in order to maintain a long shelf life and ensure quality. A large portion of fresh fish is water (*e.g.*, oysters are more than 80 percent water). Because the water in fish contains many dissolved substances, it does not uniformly freeze at the freezing point of pure water. Instead, the free water in fish freezes over a wide range, beginning at approximately -2°C. The amount of remaining free water decreases until the product reaches a temperature of approximately -40°C. Fish held below that temperature and packaged so as not to allow water loss through sublimation can be stored for an indefinite period. Unfortunately, there are relatively few commercial freezers capable of storing fish at -40° because of the tremendous variation in energy costs. Fish are therefore normally stored at -18°C to -29°C, resulting in a variable shelf life ranging from a few weeks to almost one year.

Germicidal ice: Germicidal ice is prepared by adding a chemical preservative to water before freezing. These ices are uniformly distributed (eutectic) as in the case of sodium chloride or the distribution is not uniform (non-eutectic) as with sodium benzoate. Other chemicals used in germicidal ice are hypochlorites, chloromine, benzoates, propionates, hydrogen peroxide, ozone, sodium nitrate, and antibiotics.

Heating: Heat treatment can significantly alter the quality and nutritional value of fish. Fish is exposed to heat during both the cooking process and the canning process.

Cooking: Fish is cooked in order to produce changes in the texture and flavour of the product and to kill pathogenic microorganisms. Heating fish to an internal temperature above 66°C (*i.e.,* pasteurization conditions) is sufficient to kill the most resistant microorganisms. The cooking time must be closely regulated in order to prevent excessive loss of nutrients by heat degradation, oxidation, or leaching (the loss of water-soluble nutrients into the cooking liquid).

Canning: The canning process is a sterilization technique that kills microorganisms already present on the fish, prevents further microbial contamination, and inactivates degradative enzymes. In this process fish are hermetically sealed in containers and then heated to high temperatures for a given amount of time. Canned fish can be stored for several years. However, sterilization does not kill all microorganisms, and bacterial growth and gas production may occur if the products are stored at very high temperatures. Because the severe thermal conditions of canning cause the disintegration and discoloration of the flesh of many species of fish, only a few types of fish are available as canned products. The most common types are tuna, salmon, herring, sardines, and shrimp. The thermal processing does not have a detrimental effect on the high-quality protein of the fish. In addition, these species are often canned with their bones left intact. The bones become soft and edible, significantly increasing the level of calcium present in the fish product. Tuna is an exception; because of special handling considerations, the bones of tuna are removed prior to canning. Tuna is normally caught far offshore and must be frozen and held for some period of time prior to canning. During this freezing and holding period unsaturated fatty acids are oxidized, causing the tuna to become rancid. The rancidity is removed by precooking, and the bones are removed at this time in order to facilitate the cutting and preparation of the meat for canning.

Controlling water activity: Reducing the water activity of fish inhibits the growth of microorganisms and slows the chemical reactions that may be detrimental to the quality of the fish product. The control of water activity in fish is accomplished by drying, adding chemicals, or a combination of both methods.

Drying: The principal methods of drying, or dehydrating, fish are by forced-air drying, vacuum drying, or vacuum freeze-drying. Each of these methods involves adding heat to aid in the removal of water from the fish product. During the initial stages of drying, known as the constant-rate period, water is evaporated from the surface of the product and the temperature of the product remains constant. In the final stages of drying known as the falling-rate period, the temperature of the product increases, causing water to move from the interior to the surface for evaporation.

Curing: Curing reduces water activity through the addition of chemicals, such as salt, sugars, or acids. There are two main types of salt-curing used in the fish industry: dry salting and pickle-curing. In dry salting the butchered fish is split along the backbone and buried in salt (called a wet stack). Brine is drained off until the water content of the flesh is reduced to approximately 50 percent (the typical water content of fresh fish is 75 to 80 percent) and the salt content approaches 25 percent. In heavy or hard-cure salting, an additional step is taken in which warm air is forced over the surface of the fish until the water content is reduced to about 20 percent and the salt content is increased to approximately 30 percent. Most dry-salted fish products are consumed in warm, humid countries or in areas that have few means of holding products in refrigeration or cold storage. In pickle-curing, fish are preserved in airtight barrels in a strong pickle solution formed by the dissolving of salt in the body fluids. This curing method is used for fatty fish such as herring.

Smoking: Traditionally, smoking was a combination of drying and adding chemicals from the smoke to the fish, thus preserving and adding flavour to the final product. However, much of the fish smoked today is exposed to smoke just long enough to provide the desired flavour with little, if any, drying. These products, called kippered fish, have short shelf lives, even under refrigeration, since the water activity remains high enough for spoilage organisms to grow. The smoking process consists of soaking butchered fish in a 70 to 80 percent brine solution for a few hours to overnight, resulting in a 2 to 3 percent salt content in the fish. The fish are then partially dried on racks. As the brine on the surface dries, dissolved proteins produce a glossy appearance, which is one of the commercial criteria for quality. Smoking is carried out in kilns or forced-air smokehouses that expose the fish to smoke from smoldering wood or sawdust. In cold-smoking the temperature does not exceed 29°C , and the fish is not cooked during the process. Hot-smoking is more common and is designed to cook the fish as well as to smoke it.

Irradiating: Irradiation offers a means of pasteurizing or sterilizing a variety of food products. However, the use of this process has not been universally accepted throughout the food industry. Food irradiators utilize radioisotopes, such as cobalt-60 or cesium- 137 , or electron beam generators to provide a source of ionizing radiation. The irradiation of seafood has been extensively studied since the 1950s. The pasteurization of fresh fish using low-level dosages of ionizing radiation may extend the shelf life of the product up to several weeks. The sensory and nutritional characteristics of the fish are unaffected at these low levels of radiation.

13.3. SPOILAGE

The spoilage of fish (Fig. 13.1) and sea foods depends on the following factors

i) The kind of fish
ii) The condition of fish when caught
iii) The kind and extent of contamination of fish
iv) Temperature
v) Use of germicidal ice/cooling.

Fig. 13.1 Spoiled fish

The bacteria most often involved in the spoilage of sea foods are part of the natural flora of the external slime and intestinal contents. The spoilage bacteria of the genera *Micrococcus, Bacillus, Escherichia, Proteus, Serratia, Sarcina* and *Clostridium* would mostly grow at ambient atmospheric temperature, whereas *Pseudomonas, Flavobacterium, Achromobacter* are found in chilling temperature (Table 13.1).

TABLE 13.1 MICROBIAL SPOILAGE OF FISH

SPOILAGE	ORGANISMS
1. Spoilage of fish at chilling temperature	*Pseudomonas* spp, *Achromobcter Flavobacterium*
2. Spoilage at atmospheric temperature	*Bacillus, Micrococcus, Escherichia, Proteus, Serratia, Sarcina, Clostridium*
3. Musty odour and taste	*Streptomyces* spp
4. Discolouration of fish	
Yellow	*Ps. Fluorescens,* Yellow *Micrococcus*
Pink-red	*Serratia, Micrococcus, Bacillus*
Chocolate brown	Yeast
5. Off odour in fish	*Alteromonas putrefaciens*
6. Salted fish spoilage	*Aspergillus, Penicillium, Scopulariopsis*

QUESTIONS

1. What the various sources of microbial contamination in fish and sea foods?
2. What are the microorganisms found on the outer surface of fish?
3. Comment on pre-processing and final processing of fish.
4. Write a detailed account on different methods of preservation of fish.
5. Describe the use of germicidal ice in the preservation of fish.
6. Describe the use of chilling, freezing and cold storage in the preservation of fish.
7. List out various types of microbial spoilage of fish and sea foods.
8. 'Of the many processing methods used to preserve fish, only freezing can maintain the flavour and quality of fresh fish' comment.
9. What are the four basic procedures used in the final processing of fish products?
10. Comment on discolouration and off-odour in fish spoilage.

14

CONTAMINATION, PRESERVATION AND SPOILAGE OF EGGS AND POULTRY

14.1. EGGS

India is the third largest producer of eggs. Fresh eggs are gathered on automatic collection belts at the farm and stored in a cooler at about 7°C. The eggs are then delivered to a central processing plant, where they are washed, sanitized, and graded. Grading involves the sorting of eggs into size and quality categories using automated machines. Flash candling (passing the eggs over a strong light source) detects any abnormalities such as cracked eggs and eggs containing bloodspots or other defects. Higher-grade eggs have a thick, upstanding white, an oval yolk, and a clean, smooth, unbroken shell.

14.1.1. Eggs Contamination

More than 90 percent of all eggs are free of contamination at the time they are laid; contamination with *Salmonella* bacteria and with certain spoilage organisms occurs essentially afterward. They become contaminated by faecal matter from the hen, by the lining of nest, by handling, by wash water and packing materials. The total number of microorganisms per shell of a hen's egg was found to range from 100 to 10,000,000 with a mean of about 100,000. Proper washing and sanitizing of eggs eliminates most *Salmonella* and spoilage organisms deposited on the shell. The organism *Salmonella enteritidis*, a common cause of gastroenteritis (a form of food poisoning), has been found to be transferred through the hen ovary in fewer than 1 percent of all eggs produced. Ovarian- transferred *S. enteritidis* can be controlled by thorough cooking of eggs (*i.e.*, until there are no runny whites or yolk).

14.1.2. Preservation

Methods for preservation eggs have received considerable attention because they are so perishable. Hen's egg is well protected by three structures, outer waxy membrane, shell and inner membrane.

The barriers are only temporary and offer no protection against microbial invasion through shell and membrane pores. Internally the egg white, albumin is not susceptible to microbial spoilage because of its high pH of about 9.3 and presence of antimicrobial agents (lysozyme). In contrast the egg yolk is an excellent growth medium for most of the microorganisms because of its neutral pH and nutrient content.

1. **WASHING:** Eggs are preserved by dry cleaning or washing with warm plain water that removes dirt, the bloom and part of the microorganisms, but encourages the penetration of bacteria into the eggs through the pores in the shell. A warm washing solution is essential to prevent the liquid from being drawn in to the cool egg.
2. **HEAT TREATMENT:** Heat treatments suggested include heating eggs in oil for 10 minutes at 60°C or in water for 30 minutes at 54.4°C. A thermostabilization method of dipping eggs in to hot water reduces evaporation of moisture from the eggs. Heat treatments employed for the elimination of *Salmonella* are as follows
    1. Whole egg liquid at 60°C for 3.5 minutes
    2. Plain yolk liquid at 61.1°C for 3.5 minutes
    3. Yolk with carbohydrate at 63.3°C for 3.5 minutes
    4. Yolk with salt at 63.3°C for 3.5 minutes.
3. **CHILLING:** Most shell eggs are preserved by chilling. For commercial storage for six months a temperature of - 1.7°C to - 0.5°C and relative humidity of 70% are recommended. The frozen eggs are stored at -17.8 to -20°C.
4. **USE OF PRESERVATIVES:** The preservatives may be used on the shell of eggs, in the atmosphere around them or on containers for eggs. The preservatives keep the eggs dry and reduce penetration of oxygen into the eggs and passage of carbon dioxide and moisture out. e.g. waxing, oiling, sealing. Materials used for dry packing of eggs include salt, lime, sand, saw dust and ashes. Immersion in water glass, a solution of sodium silicate is a successful method of preservation. The solution is inhibitory because of its alkalinity. Other inhibitory chemicals used are borates, permanganates, benzoates, salicylates and formats.

PRESERVATION OF EGG PRODUCTS

Liquid egg products: Refrigerated liquid egg products have become increasingly popular, especially in food-service establishments.

Liquid egg products may be delivered in a variety of packages, including bulk tank trucks, smaller portable tanks or paper cartons, hermetically sealed polyethylene bags, lacquer-coated tins, and plastic pails. These products include liquid egg whites, liquid egg yolks, and various blends of the whites and yolks. Normally, liquid egg products are pasteurized at 60°C for 3.5 minutes and have a shelf life of two to six days. Some liquid egg products are processed using ultra-pasteurization and aseptic packaging techniques to extend their shelf life to about six weeks.

Dried egg products: Dried or dehydrated eggs are less expensive to ship, more convenient to use, and easier to store than fresh whole eggs. Spray dryers are used to produce a high-quality egg product with foaming and emulsification properties similar to those of fresh eggs. The dehydrated eggs are packed in containers ranging from small pouches to large drums, depending on their commercial application. Several types of dried egg products are produced for various applications in the food industry (*e.g.*, cake mixes, salad dressings, pasta). These products include dried egg white solids, instant egg white solids, stabilized (glucose removed) whole egg solids, and various blends of whole egg and yolk with sugar or corn syrup. Most dried egg products have a storage life of one year when refrigerated.

Frozen egg products: Frozen egg products are often preferred as ingredients in certain food products. Salt, sugar, or corn syrup is normally added to yolks or whole eggs prior to freezing in order to prevent gelation or thickening of the products. Egg whites freeze well without any additives. Egg products are frozen at - 23°C and are packed in different-sized pouches and waxed or plastic cartons. Products include egg whites, egg yolks, salted yolks, sugared yolks, salted whole eggs, sugared whole eggs, and various yolk and white blends with or without added sugar or salt. At frozen temperatures they have a shelf life of about one year.

Specialty egg products: A number of specialty egg products are available to both individual consumers and institutions. Commercial salad bars utilize cryogenically frozen and diced hard-cooked eggs and pickled or plain hard-cooked eggs. Several frozen, precooked egg products are available in markets, including egg pizza, scrambled eggs, omelettes, French toast, breakfast sandwiches, crepes, and quiches. Several low-cholesterol or cholesterol-free egg substitutes have been developed by replacing the egg yolk with vegetable oils, emulsifiers, stabilizers, colour, vitamins, and minerals. Fat-free egg substitutes have also been developed for commercial use.

14.1.3. Egg Spoilage

NON-MICROBIAL SPOILAGE: Untreated eggs loose moisture during storage and hence loose weight. The white of egg become thinner and more watery as egg ages, and the yolk membrane becomes weaker to flatten out or break the yolk. During storage the alkalinity of egg white increases from pH 7.6 to 9.5. Any marked growth of chick embryo in fertilized eggs also serves to condemn the eggs.

MICROBIAL SPOILAGE: In general more spoilage of eggs (Fig. 14.1) is caused by bacteria than by molds. Certain spoilage organisms (*e.g., Alcaligenes, Proteus, Pseudomonas,* and some molds) may produce green, pink, black, colourless, and other rots in eggs after long periods of storage (Table 14.1). However, since eggs move through market channels rapidly, the modern consumer seldom encounters spoiled eggs.

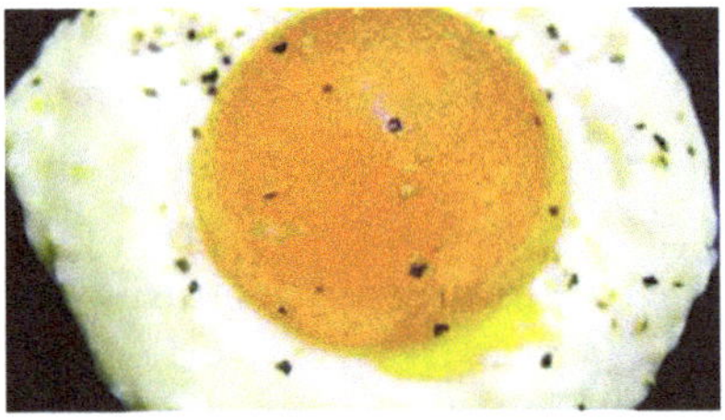

Fig. 14.1 Spoiled egg

TABLE 14.1. MICROBIAL SPOILAGE OF EGGS

SPOILAGE	SYMPTOMS	ORGANISMS
BACTERIAL ROTS		
GREEN ROT	Bright green colour of egg white in early stage, fruity odour	*Pseudomonas fluorescens*
COLOURLESS ROT	Yolk disintegrates, offensive odour	*Alcaligenes, Pseudomonas*
BLACK ROT	Yolk become blackened, putrid odour	*Aeromonas, Proteus, Pseudomonas*
PINK ROT	Pinkish precipitate on yolk	*Pseudomonas*
RED ROT	Red colour, mild odour	*Serratia*
OFF FLAVOURS	Mustiness Hay odour Fishy flavour Musty flavour	*Pseudomonas* *Enterobacter* *Escherichia coli* *Streptomyces*
MOLD ROTS		
PIN SPOT MOLDING	Small compact colonies on shell	*Penicillium* (blue spot) *Cladosporium* (black spot) *Sporotrichum* (pink spot)
WHISKERS	Superficial fungal growth	*Thamnidium, Mucor, Botrytis*

14.2. POULTRY

India is the ninth largest producer of chicken meat. Poultry is the second most widely eaten type of meat globally and, along with eggs, provides nutritionally beneficial food containing high-quality protein accompanied by a low proportion of fat. All poultry meat should be properly handled and sufficiently cooked in order to reduce the risk food contamination.

14.2.1. Contamination

The skin of live birds may contain microorganisms derived from feet, feathers and faeces. The contamination of the skin and lining of the body cavity occurs during washing, plucking and dressing. The bacterial genera generally found in poultry and poultry products include *Alcaligenes, Bacillus, Enterobacter, Escherichia, Flavobacterium, Proteus, Pseudomonas, Salmonella, Staphylococcus*. Yeasts, *Torulopsis, Rhodotorula, Candida* have been isolated from poultry and poultry products.

The incidence of Salmonellae on poultry and poultry products and its role in transmitting Salmonellosis have received considerable attention. The incidence of salmonellae positive birds have been reported to range from 0 to 20%.

14.2.2. Preservation

The method of killing, bleeding, picking or plucking the fowl has an important effect on the quality of the poultry and poultry products. The fowl is immersed in water at about 55°C for 2 min in semiscald method. Steam scalding of birds is more effective than hot water in reducing the number of bacteria, including coilforms and salmonellae. The sanitation of the housing of the birds before killing has some influence on the number of microbes on skin at dressing.

1. **CHILLING:** Most poultry and poultry products are preserved by either chilling or freezing. The dressed fowls are immersed in cold water in chill tanks. Chilling by blast of cold air would leave the birds relatively dry compared with those from chill tanks. The moisture added to the carcass may not only add organisms but also favour their growth. Chilling storage of poultry and poultry products is for a period of less than one month. The storage life is extended by 2 days at 10°C, 6 days at 4°C, 14 days at 0°C. Chickens are commercially frozen by packing in boxes lined with moisture proof and air proof paper. Most poultry is sharp frozen at -29°C in circulating air.

2. **PRESERVATIVES:** Addition of antibiotics, chlorotetracycline, or oxytetracycline to chilling bath lengthen the keeping time of poultry by several days. Soaking cut-up poultry in solutions of organic acids acetic acid , succinic acid , at pH 2.5 also lengthen shelf life. Increasing the carbon dioxide concentration in the atmosphere of stored chickens to 10-20% inhibits the growth of psychrotrophs. It is a significant factor in reducing the microbial load.
3. **SMOKING:** Poultry may be smoked. Prior to smoking, the birds must be brined (soaked in a salt solution containing certain flavourings) and then allowed to dry. Smoking can be done using real wood shavings or a smoke flavouring. In the last case this must be labeled as "natural smoke flavour added."

14.2.3. Microbial Spoilage of Poultry (Table 14.2)

TABLE 14.2 MICROBIAL SPOILAGE OF POULTRY

SPOILAGE	ORGANISM
Sour in Poultry at 10°C	*Pseudomonas,* Yeasts - *Torulopsis, Rhodotorula*
Sour in poultry above 10°C	*Micrococcus, Alcaligenes, Acinetobacter Flavobacterium, Corynebacterium*
Pigmentation	*Pseudomonas fluorescens*
Off-odour	*Alteromonas putrefaciens*

QUESTIONS

1. What the various sources of microbial contamination in eggs?
2. What are the microorganisms found on the outer surface of egg?
3. Write a detailed account on different methods of preservation of eggs.
4. Describe the use of chemicals in the preservation of eggs.
5. Describe the heat treatments employed for the elimination of Salmonella in eggs.
6. Discuss the preservation of egg products.
7. Write an essay on microbial spoilage of eggs.
8. List out various bacterial and mold rots in egg.
9. Write on non-microbial spoilage of eggs.
10. Give a detailed account on contamination, preservation and spoilage of poultry.

15

CONTAMINATION, PRESERVATION AND SPOILAGE OF MILK AND MILK PRODUCTS

15.1 CONTAMINATION

Milk is sterile at secretion in the udder, but contaminated by bacteria when it leaves the udder of a healthy cow. Milk is subject to contamination from the animal and adjacent areas. Bacteria found in manure, soil and water may enter milk from these sources. Certain diseases may be transmitted through milk, from a cow. It is essential that cows producing milk should be free from any contagious or infectious disease. Bovine tuberculosis comes from consumption of milk of tuberculous cows. Tuberculosis, brucellosis, mastitis and actinomycosis are the human disease transmitted through milk. The disease producing bacteria may find their way into milk and cause infectious diseases in children and adults.

COMMON BACTERIA IN MILK: The common bacteria found in milk are: Rickettsia, *Coxiella burnetii*, lactic acid bacteria, *Lactobacillus bulgaricus, Streptococcus thermophilus*, aerobic rods, *Bacillus subtilis, Bacillus cereus*, anaerobic bacteria, *Clostridium butyricum* .

Sources of Microbial contamination: Milk contains relatively few bacteria when it leaves the udder of a healthy cow. Milk is subject to contamination from the animal udder, skin and adjacent areas barn yard, milk utensils, milkers. Bacteria found in manure, soil and water may enter milk from these sources.

1. **UDDER:** The remnants of milk in the teat encourage microbial growth and contain bacteria such as Micrococci and Streptococci. Generally the first striping's are discarded to avoid contamination. Diseased cow and infected udder may contain pathogenic organisms such as *Actinomyces* and *Mycobacterium tuberculosis* var *bovis*.
2. **SKIN OF COW:** The microorganisms associated with soil, water, dust and faecal matter and feeds are found on the hides. The dirt,

soil and dung contain enormous number of microbes, especially Enterococci-the gas producers and putrifiers. From hides they may get deposited into milk. Dust, which gain access to milk comes from the body of the cow particularly from flanks and udder.

3. **BARN YARD:** Dust and dirt from barn yard, floor, wall and ceiling may enter milk. Spoilage organisms enter raw milk at farm level from sources like feed, the silage and grains, faecal matter and environment. Coliform are mainly associated with faecal contamination whereas clostridia come through feed and manure.
4. **MILK UTENSILS:** Milk utensils are the greatest single source of bacteria (Lactic acid bacteria) and often contribute large number of organisms to milk. Bacteria causing ropiness come from utensils and equipments.
5. **MILKERS:** Other possible sources of contamination are the hands and arm of milkman and dairy workers. The microflora of food handlers reflects the environment and habits of individuals. They are picked up from garments, soil, water, air dust. The contaminant coming from food handlers is of serious concern as it comprises of mainly of pathogenic microorganisms. The pathogenic microbes are found in infections of body surfaces - hands and arms, nasal and throat discharges of sick food handlers. *Micrococcus* and *Staphylococcus* are generally found specially associated with hands, nasal cavities and mouth. *Salmonella* and *shigella* are intestinals deposited to milk by milk handlers if sanitary practices are not followed. Typhoid fever, diarrhea, diphtheria are transmitted through milk from food handlers.
6. **WATER:** Water may contain large number of bacteria and using contaminated water for washing and rinsing the dairy equipments introduce organisms in to milk and milk products. Watering milk is an adulteration as it dilutes the milk and reduces the quality of milk supplied to the consumers. Many different kinds of microbes are found in water- bacteria, fungi, protozoa and viruses, pathogens and as well as harmless water microflora like psychrotrophs. Water-borne pathogens are *E-coli* 0157-H7, *Vibrio cholerae, Salmonella* spp.,

15.2. PRESERVATION

Milk is such a delicately flavoured, easily changed food that many preservative methods cannot be used without causing an undesirable change. In fact, most of the products made from milk evolved for the purpose of improving the keeping quality. The prevention of contamination of milk is important in the preservation. Keeping quality of milk usually improved when small number of microorganisms are present, especially those which readily grow in milk. The bacterial count of milk is used to measure its sanitary quality and milk is graded on the basis of bacterial numbers. The lactic acid bacteria and coliforms are the two major groups of microorganisms that grow well in milk. Other undesirable groups of microorganisms are the psychrotrophs, those grow well in refrigeration temperature, thermophiles, those survive the pasteurization treatment of milk and human pathogens.

1. **PASTEURIZATION:** Pasteurization of milk, widely practiced in several countries, requires temperatures of about 63°C (145°F) maintained for 30 minutes - HOLDING METHOD /LTLT or, alternatively, heating to a higher temperature, 72°C (162°F), and holding for 15 seconds-FLASH METHOD/ HTST. The times and temperatures are those determined to be necessary to destroy the *Mycobacterium tuberculosis* and other more heat-resistant of the non-spore-forming, disease-causing microorganisms found in milk. The treatment also destroys most of the microorganisms that cause spoilage and so prolongs the storage time of food.

Pasteurization is most important in all dairy processing. It is the biological safeguard which ensures that all potential pathogens are destroyed. Extensive studies have determined that heating milk to 63°C (145°F) for 30 minutes or 72°C (161°F) for 15 seconds kills the most resistant harmful bacteria. In actual practice these temperatures and times are exceeded, thereby not only ensuring safety but also extending shelf life.

Most milk today is pasteurized by the continuous high-temperature short-time (HTST) method (72°C for 15 seconds or above). The HTST method is conducted in a series of stainless steel plates and tubes, with the hot pasteurized milk on one side of the plate being cooled by the incoming raw milk on the other side. This "regeneration" can be more than 90 percent efficient and greatly reduces the cost of heating and cooling. There are many fail-safe controls on an approved pasteurizer system to ensure that all milk is completely heated for the full time and temperature requirement. If the monitoring instruments detect that something is wrong, an automatic flow diversion valve will prevent the milk from moving on to the next processing stage. Higher temperatures and sometimes longer holding times are required for the pasteurization of milk or cream with a high fat or sugar content.

Ultrapasteurized milk and cream are heated to at least 138°C for at least two seconds, but because of less stringent packaging they must be refrigerated. Shelf life is extended to 60–90 days. After opening, spoilage times for both UHT and ultrapasteurized products are similar to those of conventionally pasteurized products. Ultra-high-temperature (UHT) pasteurization involves heating milk or cream to 138 to 150°C for one or two seconds. Packaged in sterile, hermetically sealed containers, UHT milk may be stored without refrigeration for months.

Ultrahigh-temperature (UHT) pasteurization is the same heating process as ultrapasteurization (138°C for two seconds), but the milk then goes into a more substantial container-either a sterile five-layer laminated "box" or a metal can. This milk can be stored without refrigeration and has a shelf life of six months to a year. Products handled in this manner do not taste as fresh, but they are useful as an emergency supply or when refrigeration is not available.

Pasteurized milk is not sterile and is expected to contain small numbers of harmless bacteria. Therefore, the milk must be immediately cooled to below 4.4°C and protected from any outside contamination. The shelf life for high-quality pasteurized milk is about 14 days when properly refrigerated. Extended shelf life can be achieved through ultrapasteurization. Ultrapasteurized milk and cream must be refrigerated and will last at least 45 days. This process does minimal damage to the flavour and extends the shelf life of slow-selling products such as cream, eggnog (prepared using egg, milk, sugar and flavouring), and lactose-reduced milks. Pasteurization of some solid foods involves a mild heat treatment, the exact definition of which depends on the food. Radiation pasteurization refers to the application of small amounts of beta or gamma rays to foods to increase their storage time.

2. **DRYING - DRIED MILK:** Milk and by-products of milk production are often dried to reduce weight, to aid in shipping, to extend shelf life, and to provide a more useful form as an ingredient for other foods. In addition to skim and whole milk, a variety of useful dairy products are dried, including buttermilk, malted milk, instant breakfast, sweet cream, sour cream, butter powder, ice cream mix, cheese whey, coffee creamer, dehydrated cheese products, lactose, and caesinates. Many drying plants are built in conjunction with a butter-churning plant. These plants utilize the skim milk generated from the separated cream and the buttermilk produced from churning the butter. Most products are dried to less than 4 percent moisture to prevent bacterial growth and spoilage. However, products containing fat lose their freshness rather quickly owing to the oxidation of fatty acids, leading to rancidity. Two types of dryers

are used in the production of dried milk products-drum dryers and spray dryers. Each dryer has certain advantages.

Whole, low-fat, and skim milks, as well as whey and other dairy liquids, can be efficiently concentrated by the removal of water, using heat under vacuum. Since reducing atmospheric pressure lowers the temperature at which liquids boil, the water in milk is evaporated without imparting a cooked flavour. Water can also be removed by ultrafiltration and reverse osmosis, but this membrane technology is more expensive. Usually about 60 percent of the water is removed, which reduces storage space and shipping costs. Whole milk, when concentrated, usually contains 7.5 percent milk fat and 25.5 percent total milk solids. Skim milk can be condensed to approximately 20 to 40 percent solids, depending on the buyer's needs. Steam under pressure: Evaporated milk is canned and then heat processed by steam under pressure. The fore warming of milk at 100°C before evaporation kills all but more resistant bacterial spores. Sealed cans of evaporated milk are processed at 118°C for 18 minutes, which results in commercially sterile product.

3. **LOW TEMPERATURE:** Pasteurized milk is cooled to 7 to 10°C and kept that cold until processed. Ice cream and other frozen dairy products are frozen as a part of manufacturing process and stored at low temperatures in frozen state, where microbial multiplication is impossible. Freezing kills relatively few microorganisms and storage in frozen state permits survival of most of the microorganisms for long periods. Shelf life of dairy products under various storage conditions (Table 15.1).

TABLE 15.1 STORAGE CONDITIONS AND SHELF LIFE OF DAIRY PRODUCTS

PRODUCT	TEMPERATURE °C	RH %	SHELF LIFE
CHEESE (Swiss, Cheddar, Blue)	0-1.0	70	6-12 months
MILK HTST	1-4		1-2 weeks
MILK UHT	1-4		+ 1 month
Condensed MILK	0		+ 1 year
	4		6-12 months
	10		Few months
	20		few weeks
NON FAT DRY MILK	4	60	10 months

15.3. SPOILAGE

Milk is an excellent culture medium for many kinds of microorganisms, being high in moisture, neutral pH, and rich in microbial foods. A plentiful supply of food for energy is present in the form of milk sugar (lactose), butter fat, citrate, and nitrogenous compounds (proteins, amino acids, ammonia, urea, etc.) and accessory foods and minerals required by microorganisms are available. Some inhibitory substances (lactoperoxidase, agglutinins) are present in freshly drawn milk but soon become comparatively ineffective. As soon as the milk is drawn from the cow, the biochemical reactions of bacteria present in milk are initiated leading to spoilage. Bacteria are commonly associated with milk spoilage (Fig. 15.1).

Bacteria commonly associated with milk spoilage

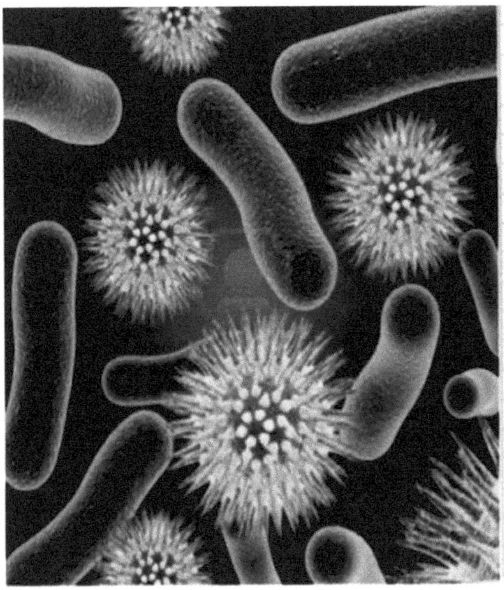

Pseudomonas	Spoilage
Brucella	Pathogenic
Enterobacteriaceae	Pathogenic and spoilage
Staphylococci	
Staphylococcus aureus	Pathogenic
Streptococcus	
S. agalactiae	Pathogenic
S. thermophilus	Acid fermentation
S. lactis	Acid fermentation
S. lactis-diacetyllatic	Flavour production
S. cremoris	Acid fermentation
Leuconostoc lactis	Acid fermentation
Bacillus cereus	Spoilage
Lactobacillus	
L. lactis	Acid production
L. bulgaricus	Acid production
L. acidophilus	Acid production
Propionibacterium	Acid production
Mycobacterium tuberculosis	Pathogenic

Fig. 15.1 Milk spoilage bacteria

TYPES OF MICROBIAL SPOILAGE

1. **Natural souring/curdling:** If the raw milk is held under ambient conditions, the immediate effect noticed is souring followed by curdling. This is primarily due to the production of acidity-production of lactic acid from lactose, by certain groups of bacteria already present in raw milk. This is the most commonly encountered

spoilage under tropical Indian conditions. The causative organisms are lactic streptococci- *Streptococcus lactis, Streptococcus cremoris, Streptococcus thermophilus, Streptococcus liquefaciens,* lactobacilli– *Lactobacillus acidophilus,* leuconostocs-*Leuconostoc dextranicum, Leuconostoc citrovorum, Bacillus coagulans* and colifoms.

2. **Gas production:** The production of gas mainly CO_2 by certain microbes is responsible for a defect called 'Gassiness'. The associative action of acid producing bacteria and gas producing yeasts cause the defect called 'Frothiness'. The production of gas in canned dairy products cause bulging of cans and the defect is called blowing of cans. The causative organisms are lactose fermenting yeasts- *Candida pseudotropicalis,Torulopsis sphaerica,* coliforms-*E.coli, Enterobacter aerogenes,* anaerobic spore forming bacteria-*Clostridium butyricum, Clostridium pasteurianum.*

3. **Ropiness/Sliminess:** Ropy fermentation of milk and cream is brought about by the growth of bacteria leading to change in consistency of the product that forms threads or viscous masses when poured. The causative organisms are gram negative rods- *Alcaligenes viscolactis, Achromobacter, Flavobacterium,* coli- aerogenes group- *Enterobacter aerogenes, Citrobacter freundii, Serratia marcescens,* micrococci-*Micrococcus viscous.* The gums (polysaccharides) derived from lactose are responsible for ropiness caused by coliforms. Mucins, which are combinations of proteins with a carbohydrate, are responsible for ropiness caused by gram negative rods and micrococci.

4. **Proteolysis:** It is the process by which casein and complex milk proteins are broken down to water soluble compounds through the action of proteolytic enzymes elaborated by microbes. The proteolysis is often accompanied by development of bitter flavour due to the presence of protein decomposition products like peptides. The causative organisms are psychrotrophic bacteria-*Pseudomonas putrefaciens, P.putida, Achromobacter, Acinetobacter, Aeromonas, Enterobacter.*

5. **Sweet curdling:** Sweet curdling is a defect associated with milk and cream, the dairy product become solid, although no acid has been produced. The defect is due to the production of an extra cellular enzyme- similar to rennin by bacteria, which causes casein to precipitate in the form of small specks of curd. The causative organisms are aerobic spore formers- *Bacillus cereus, B. subtilis, B.mycoides,* Psychrotrophic spore formers - *B. licheniformis,B. albolactis.*

6. **Lipolysis:** Lipolysis is the hydrolysis of milk fat by lipase resulting in accumulation of free fatty acids. Lipolysis of milk fat and the subsequent production of off-flavours in dairy products is a potentially important problem in cream and butter as it affects the making of the products. The lower chain fatty acids, butyric and caproic, are responsible for off-flavour, also referred as rancidity. The multiple causative factors are the intrinsic milk lipase, lipolytic organisms and their enzymes. The causative organisms for lipolysis in milk cream and cheese are psychrotrophic bacteria - *Pseudomonas fragi, Achromobacter lipolyticum, Alcaligenes viscolactis, Flavobacterium, Acinetobacter.* Yeasts and molds are considered to be the major spoilage causing organisms of cultured butter- *Candida lipolytica, C. pseudotropicalis, Geotrichum candidum, Penicillium spp, Aspergillus spp.*

7. **Off-flavours and Discolouration:** The microbially induced off-flavours such as malty flavour defect in raw milk is due to the metabolic activity of *Streptococcus lactis var maltigenes.* The malty aroma was chiefly due to the production of 3-methyl butanol from leucine. The fruity flavour-straw berry like aroma, in milk and butter held at low temperature is due to the metabolic activity and the production of ethyl esters by *Pseudomonas fragi.* The musty potato aroma in milk and cream is due to the metabolic activity of *Pseudomonas taetrolens and Pseudomonas mucidolens.* The colour of food is important aspect of their marketability. The foods are not liked by the consumers unless the colour is analogous to that naturally associated with the product. The blue colour of milk is due to *Pseudomonas syncyanca,* yellow colour due to *Pseudomonas synxantha,* red colour due to *Serratia marcescens* and green colour due to *Pseudomonas fluorescens.*

TABLE 15.2 MICROBIAL SPOILAGE OF MILK

SPOILAGE	ORGANISMS
1. SOURING OF RAW MILK -	*Streptococcus lactis, Coliforms,*
Acid production AT 10-37°C	*Enterobacter, Lactobacillus, Micrococcus*
AT 37-50°C	*Streptococcus thermophilus*
	Streptococcus faecalis
	Lactobacillus bulgaris
	Lactobacillus thermophilus
2. GAS PRODUCTION-	
foam at top of milk	*Clostridium,* Coliform bacteria
Both CO_2 and H_2	Gas forming *Bacillus*

Only CO_2	Yeasts, *Propionibcterium* Hetero fermentative lactics
3. PROTEOLYSIS	
Acid proteolysis (shrunken curd)	*Streptococcus faecalis var. liqufaciens* *Micrococcus, Enterococcus*
Alkaline proteolysis (Sweet curdling)	*Pseudomonas, Proteus* *Alcaligenes, Flavobacterium*
4. ROPINESS (sliminess)	*Alcaligenes viscolactis*
5. LIPOLYSIS (rancidity)	*Pseudomonas, Proteus* *Alcaligenes, Bacillus*
6. FLAVOUR CHANGES Sour flavour	*Leuconostoc*
Bitter flavour	Coliforms, Yeasts
Caramel flavour	*Streptococcus faecalis var. maltigenes*
7. COLOUR CHANGES -Blue milk	*Pseudomonas syncyanea*
Yellow milk	*Pseudomonas syncxantha*
Red milk	*Serratia marcescens*
Brown milk	*Pseudomonas putrefaciens*

QUESTIONS

1. What the various sources of microbial contamination in milk?
2. Write a detailed account on different methods of preservation of milk.
3. Describe the use of various methods of pasteurization in the preservation of milk.
4. Write on preservation of dried milk.
5. Discuss the storage conditions and shelf life of dairy products.
6. Write an essay on microbial spoilage of milk and milk products
7. Give a detailed account on contamination, preservation and spoilage of milk and milk products
8. Comment on Sweet curdling of milk.
9. Write on natural souring and gas production in milk.
10. Write short notes on the following - Pasteurization of milk, Ropiness in Milk, Sour milk, Colour changes in milk, Flavour changes in milk
11. Which one of the following is not a fermented milk product?

    a) Yoghurt b) Butter milk

    c) Kefir d) Skim milk

16

CONTAMINATION, PRESERVATION AND SPOILAGE OF CANNED FOODS AND BOTTLED BEVERAGES

16.1. CANNED FOODS-PRESERVATION

Canning is a method of preserving food from spoilage by storing it in containers that are hermetically sealed and then sterilized by heat. The process was invented after prolonged research by Nicolas Appert of France in 1809, in response to a call by his government for a means of preserving food for army and navy use. Appert's method consisted of tightly sealing food inside a bottle or jar, heating it to a certain temperature, and maintaining the heat for a certain period, after which the container was kept sealed until use. It was 50 years before Louis Pasteur was able to explain why the food so treated did not spoil: the heat killed the microorganisms in the food, and the sealing kept other microorganisms from entering the jar. Samuel C. Prescott and William Underwood of the United States set canning on a scientific basis by describing specific time-temperature heating requirements for sterilizing canned foods.

Most vegetables, fruits, meat and dairy products, and processed foods are stored in tin cans, but soft drinks and many other beverages are now commonly stored in aluminum cans, which are lighter and do not rust. Aluminum cans are made by impact extrusion; the body of the can is punched out in one piece from a single aluminum sheet by a stamping die. This seamless piece, which has a rounded bottom, is then capped with a second piece as its lid. The tabs used in pop- top cans are also made of aluminum. Bimetal cans are made of aluminum bodies and steel lids.

Canneries are usually located close to the growing areas of the product to be packed, since it is desirable to can foods as quickly as possible after harvesting. The canning process itself consists of several stages: cleaning and further preparing the raw food material; blanching it; filling the containers, usually under a vacuum; closing and sealing the containers; sterilizing the canned products; and labeling and warehousing the finished goods. Cleaning usually involves passing the raw food through tanks of water or under high- pressure water sprays, after which vegetable or other products are cut, peeled, cored, sliced, graded, soaked, pureed, and so on. Almost all vegetables and some fruits require blanching by immersion in hot water or steam; this process softens the vegetable tissues and makes them pliable enough to be packed tightly, while also serving to inactivate enzymes that can cause undesirable changes in the food before canning. Blanching also serves as an additional or final cleansing operation.

The filling of cans is done automatically by machines; cans are filled with solid contents and, in many cases, with an accompanying liquid (often brine or syrup) in order to replace as much of the air in the can as possible. The filled cans are then passed through a hot-water or steam bath in an exhaust box; this heating expands the food and drives out the remaining air; thus, after sealing, heat sterilizing, and cooling the can, the contraction of the contents produces a partial vacuum within the container. Certain products are vacuum-packed, whereby the cans are mechanically exhausted by specially designed vacuum- can sealing machines.

Immediately after the cans are exhausted, they are closed and sealed; a machine places the cover on the can, and the curl on the can cover and the flange on the can body are rolled into position and then flattened together. The thin layer of sealing compound originally present in the rim of the cover is dispersed between the layers of metal to ensure a hermetic seal. The sealed cans are then sterilized; *i.e.,* they are heated at temperatures high enough and for a long enough time to destroy all microorganisms (bacteria, molds, yeasts) that might still be present in the food contents. The heating is done in high-pressure steam kettles, or cookers, usually using temperatures around 240°F (116°C). The cans are then cooled in cold water or air, after which they are labeled.

16.2. CONTAMINATION

During processing of canned foods, the hot cans (120°C) are sealed and then immersed in cold water (30°C). During this step, water is sucked into the cans through minute pores in the can, leading to recontamination. The per cent of recontamination of cans was found to depend upon the microbiological quality of water (the number of bacteria present in per ml of cold water) used for cooling the cans (Table 16.1). Chlorination of water was the control applied to reduce bacterial population.

Table 16.1 RELATIONSHIP BETWEEN MICROBIAL QUALITY OF WATER AND RECONTAMINATION OF CANS

BACTERIAL POPULATION IN WATERPER ML	PER CENT OF CANS RE-CONTAMINATED
10^2	2
10^3	8
10^4	18
10^5	30
10^6	48
10^7	62

16.3. SPOILAGE OF CANNED FOODS

Most spoilage of commercially heat processed canned food resulting from under processing is caused by thermophilic bacteria, because their spores are more heat resistant than mesophilic bacteria (Fig.16.1). The five chief types of spoilage by thermophilic bacteria are

i) Flat sour spoilage
ii) TA spoilage (Thermophilic anaerobic spoilage)
iii) Sulphide spoilage
iv) Springer
v) Breather

Fig. 16.1 Spoiled canned foods

i) **Flat sour spoilage:** The ends of canned food remain flat, have a normal concavity, during souring by flat sour bacteria. Because the can retains a normal outward appearance, this type of spoilage cannot be detected by the examination of un-opened can. The spoilage occurs in low acid foods (pea or corn), is caused by *Bacillus* sp. The flat sour spoilage of acid foods (tomato juice) is caused by thermophilic species, *Bacillus coagulans. Bacillus coagulans* is a species of lactic acid forming *Bacillus* bacteria, which can contaminate canned acid foods and give a flat sour taste. It can grow at a pH as low as 4.2 and the optimum temperature is 37°C. Various species of *Bacillus* that are able to form acid without gas may be mesophilic, facultative thermophilic or obligate thermopiles like *Bacillus stearothermophilus.*

ii) **TA spoilage (Thermophilic anaerobic spoilage):** This spoilage is caused by thermophilic anaerobic bacteria like *Clostridium thermosaccharolyticum,* which form acid and gas (carbon dioxide) in low acid and medium acid foods. They do not produce hydrogen sulphide.

iii) **Sulphide spoilage:** The bluish gray liquid in canned corn food is evident of the spoilage and caused by *Clostridium nigrificans.* It is detected by means of black (FeS) colonies it forms in an iron sulphite agar at 55°C. Sulphur stinker is caused by *Desulfomaculum nigricans,* an obligate thermophile found in low acid canned foods of peas and corns due to under processing in low temperature with poor cooling. The sources of the spoilage organism are manure and plant equipments. The spoilage is evident by hydrogen sulphide odour and bluish grey liquid.

iv) **Springer:** The container appears bulged in one end of the can or both the ends. The bulging resembles as if the container is dashed against a hard object. This is due to the production of acid as well as gas by the bacteria causing this spoilage.

v) **Breather:** The ends of the containers are pushed inside and it is likely to take atmospheric air inside, if holes are found. This type of spoilage may be due to improper packing or microbial action.

TABLE 16.2. MICROBIAL SPOILGAE OF CANNED FOODS

SPOILAGE	FOODS	ORGANISMS
1. Spore Forming Mesophilic Spoilage	Canned or home canned acid foods-Pine apple, Tomato Protein foods-proteolytic	*Clostridium pasteurianum* *Clostridium butyricum* *Clostridium sporogenes* *Clostridium putrefaciens* *Clostridium botulinum*
	Sea foods, meat, condensed milk	*Bacillus subtilis, Bacillus mesenteroides*
	Canned peas, tomato, peaches	*Bacillus polymyxa, Bacillus macerans*
2. Non-spore Forming Bacterial Spoilage	Under processed fruits and vegetables	*Streptococcus thermophilus Micrococcus, Lactobacillus, Leuconostoc, Pseudomonas*
3. Yeast Spoilage	Canned fruits, jams, jellies, Fruit juices, syrups	Film yeast
4. Mold Spoilage	Home canned jam, jelly	*Penicillium*

16.4. BOTTLED BEVERAGES

SOFT DRINKS: Any of a class of nonalcoholic beverage, usually but not necessarily carbonated, normally containing a natural or artificial sweetening agent, edible acids, natural or artificial flavours, and sometimes juice are called soft drinks. Natural flavours are derived from fruits, nuts, berries, roots, herbs, and other plant sources. Coffee, tea, milk, cocoa, and undiluted fruit and vegetable juices are not considered soft drinks.

The term was originated to distinguish the flavoured drinks from hard liquor, or spirits. Soft drinks were recommended as a substitute in the effort to change the hard-drinking habits of early Americans. Indeed, health concerns of modern consumers have led to new categories of soft drinks emphasizing low caloric and sodium content, no caffeine, and "all natural" ingredients.

There are many specialty soft drinks. Mineral waters are very popular in Europe and Latin America. Kava, made from roots of a bushy shrub, *Piper methysticum*, is consumed by the people of Fiji and other Pacific islands. In Cuba people enjoy a carbonated cane juice; its flavour comes from unrefined syrup. In tropical areas, where diets frequently lack sufficient protein, soft drinks containing soybean flour have been marketed. In Egypt carob or locust bean extract is used. In

Brazil a soft drink is made using mate as a base. The whey obtained from making buffalo cheese is carbonated and consumed as a soft drink in North Africa. Some Eastern Europeans enjoy a drink prepared from fermented stale bread. Honey and orange juice go into a popular drink of Israel.

The first iced soft drink consisted of a cup of ice covered with a flavoured syrup. Sophisticated dispensing machines now blend measured quantities of syrup with carbonated or plain water to make the finished beverage. To obtain the soft ice, or slush, the machine reduces the beverage temperature to between -5°C and -2°C.

PRESERVATION: Packaging and vending: Soft drinks are packaged in glass or plastic bottles, tin-free steel, aluminum, or plastic cans, treated cardboard cartons, foil pouches, or in large stainless steel containers. Vending of soft drinks had its modest beginning with the use of ice coolers in the early 20th century. Nowadays, most drinks are cooled by electric refrigeration for consumption on the premises. Vending machines dispense soft drinks in cups, cans, or bottles, and restaurants, bars, and hotels use dispensing guns to handle large volume. There are two methods of vending soft drinks in cups. In the "pre-mix" system, the finished beverage is prepared by the soft drink manufacturer and filled into five- or 10-gallon stainless steel tanks. The tanks of beverage are attached to the vending machine where the beverage is cooled and dispensed. In the "post-mix" system the vending machine has its own water and carbon dioxide supply. The water is carbonated as required and is mixed

The carbonation in soft drinks is inhibitory or even germicidal to some microorganisms but not to others. The acidity in soft drinks resulting from carbonation and the addition of acids (citric, lactic) inhibits the growth of microorganisms not tolerant to acidity. Non-acid drinks, such as root beer, are better culture media for spoilage microorganisms than acid drinks such as cola, ginger ale, and fruit flavour drinks.

SPOILAGE: Yeasts chiefly *Torulopsis, Candida* are the most likely causes of spoilage of soft drinks. Cloudiness or ropiness are types of spoilage of soft drinks Cloudiness results from marked growth of yeasts or bacteria and ropiness results from the development of capsulated bacteria of the genus *Bacillus*.

QUESTIONS

1. Write a brief account on invention and development of canning.
2. Explain the stages in canning.
3. Write a brief account on contamination of canned foods
4. What are the chief types of spoilage of canned foods by thermophilic bacteria?
5. List out various kinds of spoilage of canned foods and the causative organisms
6. Write an account on microbial spoilage of canned foods
7. Explain how bottled drinks are preserved.
8. Write a brief account on microbial spoilage of soft drinks.
9. Write short notes on

   i) TA spoilage

   ii) Sulphide spoilage Processing of canned foods
10. Flat sour in canned food is caused by

    a) *Bacillus coagulans* b) *Clostridium botulinum*

    c) *E.coli.* d) *Clostridium nigrificans*
11. *Clostridium nigrificans* causes this canned food spoilage

    a) Flat sour b) Springer

    c) Breather d) Sulphide spoilage

17

FOOD POISONING

Food poisoning is an acute illness resulting from the ingestion of injurious agents such as chemical, heavy metals, parasites, microbial toxin, wild species of mushroom in food. Food poisoning is commonly caused by certain bacteria or their toxins, which are the poisonous proteins that these bacteria produce. Food borne pathogens are potential global problem, more than 90 per cent of the cases of food poisoning are caused by bacteria, *Staphylococcus aureus, Salmonella* spp, *Clostridium perfringens, Campylobacter jejuni, Listeria monocytogenes, Vibrio parahaemolyticus, Shigella dysenteriae, Bacillus cereus* and enteropathogenic *E. coli*. These bacteria are commonly found in many raw foods. Generally a large number of food poisoning bacteria must be present to cause illness. In most of these food poisonings, the food serves as a vehicle of transmission. The role of food here is significant, since the product not only may permit the survival of the pathogens, but also may provide a suitable medium for the rapid multiplication of microbes and production of toxin. A more serious aspect of microbial contamination is the possibility of serious health hazards due to potential pathogenic organisms gaining entry into foods due to low standards of hygiene and sanitation.

Food poisoning may be due to food intoxication or food infection. Food intoxication and food infection differ in many characteristics (Table 17.1). Food intoxication refers to food borne illness caused by the ingestion of a microbial toxin preformed in the ingested food (Botulism). The most serious food poisoning caused by toxins is botulism, an often fatal disease that results from eating improperly canned foods containing the bacteria *Clostridium botulinum*. Classically, food intoxication is caused by the growth of *Clostridium botulinum, Staphylococcus aureus, Bacillus cereus* in food. *Clostridium botulinum* produces a neurotoxin, while *Staphylococcus aureus* and *Bacillus cereus* cultures produce enterotoxins in food.

TABLE 17.1. DIFFERENCES BETWEEN FOOD INTOXICATION AND FOOD INFECTION

FOOD INTOXICATION	FOOD INFECTION
1. It is a toxic disease (Botulism).	It is a infectious disease (Salmonellosis).
2. Bacteria multiply in offending food and produce toxin before it is eaten. Manifestation comes from ingested toxin.	Bacteria that have been taken along with the food multiply in the body and cause the disease.
3. They produce and release exotoxin, a type of protein denatured by heat.	They produce endotoxin a type of carbohydrate, lipid, lipopolysacchride bound to cell wall released on cell lysis.
4. Exotoxin generally produced by Gram positive bacteria.	Endotoxin generally produced by Gram negative bacteria.
5. Exotoxins have high toxicity.	Endotoxins have low toxicity.
6. As they do not multiply in intestine, culture methods can not be used and toxins are diagnosed.	Culture methods are used to identify the pathogenic organisms.
Examples :*Clostridium botulinum, Staphylococcus aureus, Bacillus cereus.*	Examples: *Salmonella, E.coli, Shigella, Vibrio, Campylobacter, Listeria, Clostridium perfringens.*

Food infection refers to food borne illness caused by the entrance of viable, usually multiplying pathogenic bacteria into the body through the ingestion of contaminated food. e.g. Samonellosis. In some cases, food poisoning is caused not by toxins but by rapidly growing colonies of the bacteria themselves. Among the most common of these harmful bacteria are *Salmonella enteritidis* and *Salmonella typhimurium, Clostridium perfringens, Campylobacter jejuni, Listeria monocytogenes, Vibrio parahaemolyticus, Shigella dysenteriae, Yersinia enterocolitica* and enteropathogenic *E. coli.* These bacteria are commonly found in many raw foods, which may be spread to food by mice, rats, or humans-especially un-hygeinic workers in restaurants and food processing plants.

Toxins produced by food pathogens may be classified based on method of production or type of symptoms. Based on method of production, toxins are grouped as exotoxins and endotoxins. **Exotoxin** is secreted by the bacteria into surrounding medium. The toxin, denatured by heat, is a type of protein produced by gram positive bacteria and negative bacteria. The toxicity is great. e.g. Botulin toxin. **Endotoxin** is bound to cell wall and liberated only when the cell producing it disintegrates. The toxic principle is carbohydrate, lipopolysaccharide (LPS) or lipid. It is generally produced by gram negative bacteria and toxicity low. e.g. Salmonellosis toxin.

Based on symptoms, toxins are classified as enterotoxin and neurotoxin. **Enterotoxin** causes gastroenterititis, the inflammation of the lining of intestinal tract. The principal symptoms are nausea, vomiting, abdominal pain (cramps) and diarrhea. e.g. toxins produced by *Salmonella, Vibrio, Shigella, E. coli, Staphylococcus, Bacillus cereus.* **Neurotoxin** is produced by *Clostridium botulinum* affects muscles of vision, swallowing and respiration. Symptoms consist of generalized weakness, fatigue, double vision, slurred/thickness of speech, difficulty in swallowing, respiratory failure. The toxin blocks acetylcholine at neuromuscular junctions halts neuronal activity and muscles are paralyzed. The toxin also dilates blood vessels, hemorrhage appear in different parts of the body and sometimes death.

17.1. FOOD INTOXICATION

Food intoxication is caused by bacterial toxins pre-formed in food. Often the toxins produced are more stable than the bacteria themselves. The *Clostridium botulinum* produces a neurotoxin while *Staphylococcus aureus* and *Bacillus cereus* cultures produce enterotoxins.

17.1.1. Botulism

It is a very serious type of food intoxication disease caused by *Clostridium botulinum*. The disease derives its name from the Latin word 'Botulus', which means sausage. The major source of botulism is improperly processed home canned vegetable sausage of low acid content (peas, beans, asparagus sausages).

Organism: *Clostridium botulinum* is an anaerobic, Gram positive, spore forming bacillus. It is a common inhabitant of soil, hay, vegetable, sea mud, animal manure, etc, the usual sources from which foods are contaminated. The organism is primarily a saprophyte and unable to grow inside the body of a warm blooded animal. The spores are wider than the bacillary bodies giving the bacillus a swollen appearance resembling spindle, hence the name *Clostridium* (Fig. 17.1). It produces subterminal, oval or bulging endospores. It can exist as a heat-resistant spore, and can grow and produce neurotoxin in under processed home canned foods. The toxin is destroyed in 10 min at a temperature of boiling water, but the spores must be held at 120°C for 15 min to be killed. It is identified in laboratory based on the characters: motile, no terminal spore, non-aero tolerant and lipase positive.

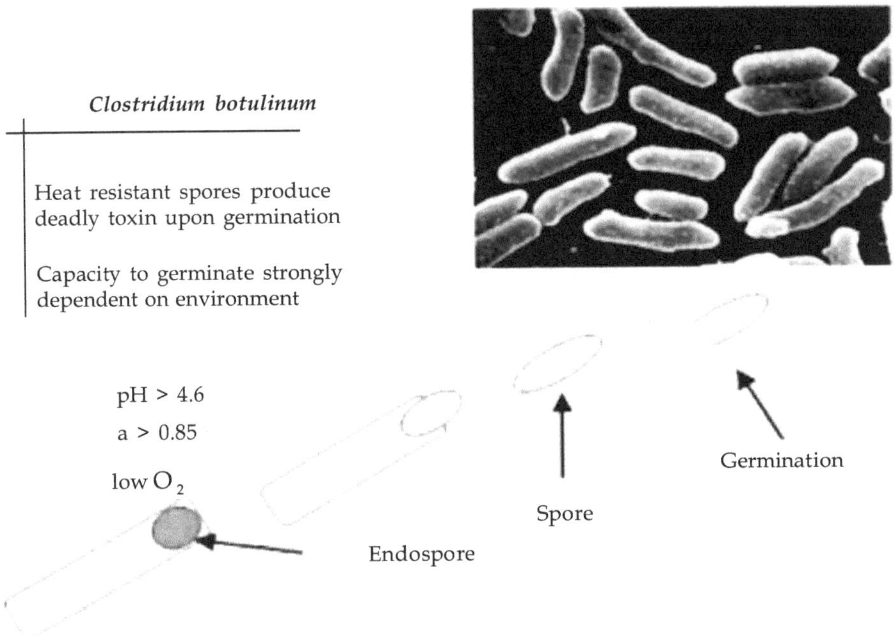

Fig. 17.1 *Clostridium botulinum*

Toxins: Botulinum toxins are extremely potent neurotoxins. It is the most deadly of poisons, the mere tasting of contaminated food having been known to cause death. The preformed toxin in food is absorbed through intestinal wall to exert its neurotoxic effect. The toxin is odourless and tasteless! The neurotoxin, NTX ~ 150 KDa, of *Clostridium botulinum* is in the form of a stable complex due to its association with non-toxic proteins. The 7S neurotoxin is synthesized as a single polypeptide chain of molecular weight 150 KDa, which undergoes cleavage to form a di-chain molecule. The heavy chain correlates with the binding of toxin to peripheral synapses and the light chain is associated with the intracellular activity of blocking acetylcholine release. A fast acting neurotoxin it affects peripheral nervous system - the toxin blocks release of acetylcholine at neuromuscular junctions, neuronal activity halts and muscles are paralyzed.

Symptoms: Toxic signs appear within 24 - 48 hrs. Muscles of vision, swallowing and respiration are affected. Symptoms consist of generalized weakness, visual disturbances (double vision), thickness of speech, nausea, vomiting and difficulty in swallowing. Toxin also dilates blood vessels and hemorrhages appear in different parts of the body. Death usually occurs between third and seventh days. The lethal dose is 1-2 microgram. The mortality rate ranges from 30 to 100%.

Toxigenic types: There are seven antigenically distinct toxigenic types of *Clostridium botulinum* namely, A,B,C.D.E.F and G. Human botulism is always result from ingestion of toxins from types : A,B,E,F and G. The C and D toxins cause diseases in animals.

1. **A and B:** These toxins are of plant origin and found in home preserved vegetables and fruits. A is more toxic than type B, while B is found more often than type A.
2. Toxin **E** is found in fish and marine food products. The spores of type E can germinate at refrigeration temperature and form toxin.
3. Toxin **F** has been reported in home processed meat called Venison jerky in Denmark, meat sliced into strips and dried in sun.
4. Type **G** has been isolated from soil contaminated foods in Argentina. It has not been implicated in human botulism.

The toxigenicity of clostridial species and type of toxin depend on the presence of specific lysogenic bacteriophages. If *Clostridium botulinum* type C is cured of its phage, it loses its toxin. A startling fact is that, experimentally the right type of phage can covert the bacterium to a different clostridial type, also toxigenic and producing that specific toxin!.

17.1.2. *Staphylococcus enterotoxicosis*

It is caused by the action of a preformed enterotoxin liberated by *Staphylococcus aureus* and *Staphylococcus epidermis.* Man's respiratory passages, skin and superficial wounds are common sources of *Staphylococcus aureus.* When *Staphylococcus aureus* is allowed to grow in foods, it can produce a toxin that causes illness. Although cooking destroys the bacteria, the toxin produced is heat stable and may not be destroyed. The toxin liberated is termed as enterotoxin because it causes gastroenteritis or inflammation of the lining of intestinal tract. It produces six serologically distinct enterotoxins, A, B, C, C_2, D and E that differ in toxicity. A and D are more often associated with food poisoning outbreaks. These enterotoxins are proteins with molecular weights between 26,000 and 30,000.

Staphylococcus aureus is Gram positive facultative anaerobic cocci, catalase positive, coagulase positive, enteropathogenic bacterium produce golden yellow pigments. Growth temperature ranges from 10 to 45°C, but their optimum is 37°C. pH range is 5 to 9 and minimum water activity required is 0.96. They are resistant to salt , can tolerate up to 15% NaCl.

Staphylococcal food poisoning occurs most often in foods that require hand preparation, such as potato salad, sandwich spreads and vegetable salads. It occurs almost everywhere and grows readily in foods stored at room temperature, especially processed meat and fish, milk, and cream-filled foods. Foods are left at room temperature for long periods of time, allowing the bacteria to grow and produce toxin in meat products and marine products. Symptoms appear in 2- 6 hours, with no fever. Sometimes the good personal hygiene of food handlers will help keep *Staphylococcus aureus* out of foods, and refrigeration of raw and cooked foods will prevent the growth of these bacteria if any are present.

17.1.3. *Bacillus cereus* Gastroenteritis

This Gram positive, facultative anaerobic/aerobic, spore forming rod bacteria are found in dust, soil and spices. Produce blue coloured rhizoidal colonies. Ferment glucose, reduce nitrate and do not ferment mannitol as carbon source. Growth temperature ranges from 30 to 35°C, but their optimum is 37°C, pH range is 4.5 to 9 and bacilli tolerate 5% NaCl.

It can survive normal cooking as a heat resistant spore, and then produce a large number of cells if the growth conditions are favourable. Considering the dual phase of *Bacillus cereus* to exist as vegetative cell and spore, thermal inactivation pattern of the toxigenic culture of *Bacillus cereus* indicated that the death rate was not constant during the process of heat treatment. Starchy foods such as rice and potato dishes, dairy products (milk, cheese), spices, pulses, snacks, lunch foods are most often involved. It causes food intoxication through enterotoxins produced in contaminated food and onset of symptoms occurs in 2 to 24 hrs. The diarrhea enterotoxin is thermolabile, denatured at 55°C in 20 minutes. The emetic (fever) enterotoxin is thermostable, denatured only at 126°C in 90 minutes.

17.2. FOOD INFECTIONS

Food borne infections are mostly caused by bacteria and viruses. In this type of food poisoning, the viable bacterial cells have been ingested with food. The dose required to produce an infection varies with the type of microbe, even though the organism usually will multiply in the gastro-intestinal tract to produce the infectious disease. Common bacteria involved in food borne infections are: *Salmonella* spp., *Clostridium perfringens, Vibrio parahaemolyticus, Campylobacter jejuni, Listeria monocytogenes, Escherichia coli, Shigella dysenteriae, Yersinia enterocolitica, Helicobacter pylori.* The three food pathogens-*Yersinia enterocolitica, Listeria monocytogenes, Campylobacter jejuni* are involved in large number of outbreaks in recent period and termed as "Emerging pathogens of public health significance". This has been possible due to their tolerance to adverse conditions, survival and multiplication in refrigerated foods.

17.2.1. Salmonellosis

Salmonellosis may result following the ingestion of viable cells of *Salmonella* spp. There are 250 species and more than 2000 serovars of *Salmonella*. It is the most frequently occurring bacterial food infection. An American vetnary pathologist, Daniel Elmor Salmon who isolated it in 1885, first described it.

Organism: *Salmonella* (Fig. 17.2) are Gram negative, nonspore forming rods, motile, enteric. The most frequent species associated with food infection in humans is *Salmonella typhimurium*. It is resistant to all antibiotics. *S. choleruesuis* and *S. enteritidis* also cause salmonellosis. Growth temperature ranges from 6.7 to 45.6°C, but their optimum is 37°C. pH range is 4 to 9 and minimum water activity required is 0.93.

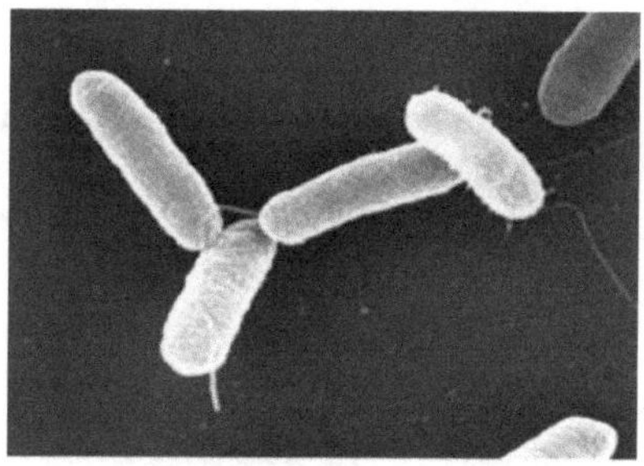

Fig. 17.2 *Salmonella*

Sources: The gastrointestinal tracts of animals and man are common sources of *Salmonella* found in foods. High protein foods such as meat, poultry, fish and eggs are most commonly associated with *Salmonella*. This bacterium is transmitted through water, milk, or food contaminated with faeces. The other sources are mice, rats, cockroaches, poultry, eggs and flies. *Salmonella* are abundant in the intestinal tract of poultry, a very important source and the contaminated eggs may be a source of the organism in egg products. **Foods involved:** contaminated meat and meat products, poultry, egg products, custard, cream, milk, salads.

Symptoms: The principal symptoms of a salmonella gastrointestinal infection are headache, nausea, vomiting, abdominal pain, and diarrhea. This may be preceded by a headache and chills. The mortality is low, being less than 1%.

Prevention: Cook thoroughly, avoid cross contamination by following sanitary practices. *Salmonella* cells are destroyed at cooking temperature above 70°C. If *Salmonella* is present in raw or cooked foods, its growth can be controlled by refrigeration below 5°C.

17.2.2 *Clostridium perfringens*

It is Gram positive, non- motile, anaerobic spore forming rod shaped bacteria found in soil, dust, and intestinal tracts of animals and man. Growth temperature ranges from 15 to 50°C, but their optimum is 37°C. pH range is 5.5 to 8 and minimum water activity required is 0.93. When food containing large number of these bacteria is consumed, the bacteria produce a toxin in intestinal tracts that causes illness. It can exist as a heat resistant spore and survive cooking. The bacteria multiply if the cooked food is held between 5°C and 60°C for a long time.

Symptoms include acute abdominal pain, diarrhea, and gas. The enterotoxin produced by the bacterium during sporulation is heat sensitive and inactivated at 65°C in ten minutes. Spores are found in raw food, soil, sewage and animal faeces. Cooked meat held for some time cool before consumption, cold chicken, fish paste, beef, pork, sauces and gravies are the foods most frequently involved in the food poisoning.

17.2.3. *Vibrio parahaemolyticus*

It is Gram negative, curved or straight capsulated. motile rod produces peritrichous flagella in solid media and polar flagella in liquid media. The halophilic bacteria that require 1-3% NaCl, found in sea foods cause the disease, VIBRIOSIS. The bacterium is very sensitive to cold and heat. The bacteria in sea foods are destroyed at below 5°C and above 60°C. Symptoms are abdominal pain, diarrhea and no fever. It is a major problem in Japan where many sea foods are consumed raw. The bacterium is responsible for 60% cases of summer diarrhea in Japan. The consumption of raw or inadequately cooked sea foods - shrimps, blue crabs, and shellfish, , calms, green mussels, oysters cause the food infection. *Vibrio vulnificus* is another member of the *Vibrio* genus that is found in marine environments (a moderate halophile) and truly an emerging pathogen. Eating sea foods contaminated with this organism lead to gastroenteritis, which usually develops within 16 hours of eating contaminated food. Wound infections result from contaminating an existing open wound with sea water harbouring the organism. They occur in both fresh and marine water habitats and in association with aquatic animals.

17.2.4 *Campylobacter jejuni*

They are Gram negative, microaerophilic, spiral bacteria, thermophilic in nature, require an optimum temperature of 42°C. They are sea gull shaped, highly motile cells, move in cork screw fashion, cause the disease, Campylobacteriosis. The organism is found in intestinal tract, reproductive organs and oral cavity of infected humans and animals. The faecal contamination of food and water and contact with sick people or animals may cause the illness. The raw milk and milk products are the foods most frequently involved. Poultry and meat products are the major reservoirs and vehicles. It is primarily transferred from animal origin foods to humans.

Symptoms include watery, bloody diarrhea. The illness is sudden. It is the most common bacterial agent of Mucosanguineous diarrhea. Other symptoms are fever, abdominal pain, nausea, headache and muscle pain.

Prevention: The principles of animal science will play significant role in control of this ubiquitous organism. Hygienic slaughter and processing procedures, and proper storage conditions will cause a decrease in the microbial load. Treatment with erythromycin reduces the infection. Chlorinating drinking water kills the bacteria.

17.2.5. *Listeria monocytogenes*

It is Gram positive, aerobic or microaerophilic bacillus occur in chains, found in animals, soil, water and sewage. This motile bacterium is now recognized as an important food pathogen. It can be isolated from soil, silage and other environmental sources. It is quite hardy and resists the deleterious effects of freezing, drying and heat remarkably well for a bacterium that does not form spores. It can grow in pH range of 5 to 9.5 , temperature 4 to 40 °C in growth medium. The organism has survived the pH 5 environment of cottage and cheddar cheese. It is salt tolerant, surviving concentrations as high as 30 percent. It can grow slowly and multiply at refrigeration temperature. Since high heat, greater than 80°C eliminates the organism, post-process contamination from environmental sources then becomes a critical control point for many foods.

Foods involved: *Listeria monocytogenes* has been associated with foods such as raw milk, pasteurized fluid milk, soft ripened cheeses, ice cream, raw vegetables, raw and cooked poultry, raw meats, raw and smoked fish (crab, lobster, shrimp). The greatest threat of listeriosis is from ready-to eat products that do not require further cooking at home. The ability to grow as low as 3°C permits its multiplication in refrigerated foods.

Ingestion of this bacteria causes listeriosis in humans marked by a flu-like illness. Immunocompromized humans such as pregnant women, cancer patients, alcoholics, elderly and newborn are highly susceptible to virulent *Listeria* and the mortality rate is approximately 30 percent among them. The manifestations of listeriosis include septicemia, meningitis, encephalitis and cervical infection in pregnant women, which may result in spontaneous abortion or still birth. Hazards from *Listeria monocytogenes* can be prevented by thoroughly cooking sea foods and preventing cross contamination of cooked foods.

17.2.6. Enteropathogenic *Escherichia coli*

E. coli are Gram negative, rod shaped, motile by peritrichate flagella, aerobe and facultative anaerobe, with growth optimum at 37° C. The coliform bacteria found in gut and helping in the control of more harmful microorganisms . Most of the strains of *E.coli* are harmless commensals, but some strains are virulent pathogens. There are four subgroups of pathogenic *E. coli,* enterotoxigenic (ETEC), enteroinvasive (EIEC), enterohaemorrhagic (EHEC) and enteropathogenic (EPEC) with distinct characteristics. (Table 17.2) ETEC strains are the primary cause of infant diarrhea and travellers' diarrhea. EIEC strains invade intestinal wall resulting in inflammation, fever and dysentery. Fluid loss is greater with invasive strains. EHEC strains of Shiga like toxin producing *E.coli* (0157:H7, 0104:H21 and 011:NM) have been recognized as human pathogens causing bloody diarrhea known as haemorrhagic colitis. In young children, haemolytic uremic syndrome is a serious complication that leads to renal failure and death. EPEC strains of *E. coli* cause diarrhea in new born infants, sometimes causing epidemics in hospital nurseries. *E-coli* 0157:H7 is a leading cause of food- borne illness. The major sources of these organisms in the environment are faeces of infected humans and untreated water. The ground beef, raw milk, unpasteurized apple juice, home-made venison jerky and untreated water are the foods most frequently involved. Adequate cooking, avoidance of recontamination of cooked foods by contaminated equipment, water or infected food handlers prevent the infection.

TABLE 17.2. ENTEROPATHOGENIC *Escherichia coli TYPES*

S.NO.	TYPE	E.COLI	CHARACTER	OCCURENCE
1.	EPEC	Enteropathogenic	Institutional diarrhea outbreaks	Diarrhea in Infants
2.	ETEC	Enterotoxigenic	Traveller's diarrhea in tropics	Diarrhea in all age groups
3.	EIEC	Enteroinvasive	Mild diarrhea to dysentery	Diarrhea in both children and adults
4.	EHEC/ STEC	Enterohemorr-hagic/Shiga like toxin producing	Food poisoning diarrhea Serotype O157:H7	Diarrhea in school children

17.2.7 Shigellosis

Shigellosis or bacterial dysentery is caused by four species of pathogenic *Shigella,* including *S. dysenteriae, S.sonnei, S.boydii* and *S. flexneri*. These organisms can persist in foods for up to a month. *Shigella* spread rapidly in overcrowded conditions with poor sanitation. The pathogenic species are spread by eggs, vegetables, dairy products, shellfish and contaminated water. The food-borne *Shigella* cells survive the acidity of stomach, pass into the gut, which is the primary site of the disease. They release *Shigella* toxins, inflammation develops around multiplying bacteria, leading to abscess and bleeding ulceration. The symptoms include abdominal cramps, fever, profuse diarrhea with mucous and blood in the excreta.

17.2.8. Yersiniosis

Yersiniosis is infection with *Yersinia enterocolitica*, short plump ovoid, Gram negative bacillus, with round ends and convex sides arranged singly or in short chains. Stained cells appear as safety pin, two ends densely stained and central area clear. Pleomorphism is enhanced in media containing 3% NaCl. Aerobic, facultatively anaerobic grow in wide pH(pH5.0-9.6, optimum pH 7.2) and temperature (range 2-45°C, optimum temperature 29 °C)motile at 22°C. *Yersinia enterocolitica* has been isolated from a wide range of domestic and wild animals and recently reported in human clinical material. It cause enterocoliticus-gastroeneritis in children, mesenteric adenitis in older children that may mimic appendicitis, systematic disease in adults characterized by bacteremia. It is generally asociated with under cooked beef and pork and milk.

FOOD BORNE VIRUSES: Hepatitis A: Chief among the viruses that cause food poisoning is hepatitis A, which is excreted in the faeces of infected individuals and re-enters the food chain through unsanitary methods of food preparation. Hepatitis A Virus (HAV) is highly contagious and transmitted from person to person from faecal-oral route and ingestion of contaminated food or water. Water, salads, shellfish, and milk products are common sources of contamination. The virus, HAV, contains a single stranded RNA in a cubic symmetry without an envelope. The enterovirus particles are 27 nm in diameter and serologically distinct.

Hepatitis A infection causes a mild illness with symptoms that include fever, fatigue, nausea, loss of appetite, soreness in the abdominal area, and jaundice (accumulation of the pigment bilirubin in the blood, which causes yellowish discoloration of the skin and whites of the eyes). HAV, may cause, infectious hepatitis, short inclusion hepatitis, catarrhal jaundice or epidemic jaundice. The incubation period is 28 days. The disease is rare in children and more often found in adults. Good personal hygiene, safe sex, proper sanitation, avoiding partially cooked and raw foods prevents the spread of Hepatitis A. The food borne hepatitis E virus (HEV) has symptoms identical to those of hepatitis A. Other viruses that cause food poisoning include the rotaviruses, which cause acute upset of the digestive system and the Norovirus (NoV) family of human enteric viruses, found in sea food samples, oysters and calms which also cause gastroenteritis.

PROTOZOA Amoebiasis (*Entamoeba histolytica*): It is an anaerobic parasitic protozoan, predominantly infecting humans and other primates. *E. histolytica*, as its name suggests (*histo-lytic* = tissue destroying), is pathogenic; infection can lead to amoebic dysentery. The active stage, trophozoite exists only in the host and in fresh loose faeces; cysts survive outside the host in water, soils and on foods under moist conditions. The cysts are readily killed by heat and by freezing temperatures, and survive for only a few months outside of the host. When cysts are swallowed they cause infections by excysting (releasing the trophozoite stage) in the digestive tract. The trophozoite stage is readily killed in the environment and cannot survive passage through the acidic stomach to cause infection. The life cycle stages of *E. histolytica* are : cysts- (excystation) progeny amoeba- trophozoites- (encystation) uninucleate cysts.

Symptoms can include fulminating dysentery, bloody diarrhea, weight loss, fatigue, abdominal pain, and amoeboma. The amoeba can actually 'bore' into the intestinal wall, causing lesions and intestinal symptoms, and it may reach the blood stream. From there, it can reach different vital organs of the human body, usually the liver, but sometimes the lungs, brain, spleen, etc. A common outcome of this invasion of tissues is a liver abscess, which can be fatal if untreated. Ingested red blood cells are sometimes seen in the amoeba cell cytoplasm.

Blantidiasis and Giardiasis are the food and water borne infections caused by protozoa, transmitted by faecal contaminated food. Blantidiasis is caused by *Blantidium coli*, a ciliated pathogenic protozoan of human intestine, the symptoms include abdominal cramping, vomiting, ulcer, dysentery and weight loss. Giardiasis caused by *Giardia lamblia*, a flagellated pathogenic protozoan of human intestine, the symptoms include nausea, weakness, chronic prolonged greasy diarrhea. The encysted protozoan is resistant to chlorine.

CIGUATERA POISONING: It is contracted by man upon ingestion of numerous species of reef or shore fish such as barracudas, groupers, sea basses, jacks and eels. They acquire poison as a consequence of feeding on smaller herbivorous fish, which in turn feed on toxic algae (like *Lyngbya majuscule*), plankton or plants. Symptoms of ciguatera poisoning develop within four hours following ingestion of offending fish and consist of nausea, paresthesia about the mouth, tongue and throat and sometimes over the face and distal parts of fingers and toes. Weakness, abdominal pain, vomiting, diarrhea and chills are the other symptoms. Important food borne diseases are listed in Table 17.3.

TABLE 17.3 IMPORTANT FOODBORNE DISEASES

DISEASE	CAUSATIVE ORGANISM	FOODS	SYMPTOMS
Botulism	*Clostridium botulinum*	Sausages, home made canned foods	Visual disturbances, thickness of speech, nausea, vomiting and difficulty in swallowing.
Staphylococcal food poisoning	*Staphylococcus aureus*	Hams, meat, poultry, cheese, salad, sandwich, high salt foods	Nausea, vomiting, diarrhea, no fever
B. cereus food poisoning	*Bacillus cereus*	Starchy foods, Meat, sauces, vegetables.	Diarrhea, cramps, vomiting.
Clostridium perfringens food-borne illness	*Clostridium perfringens*	cold chicken, fish paste, beef, pork, sauces and gravies	acute abdominal pain, diarrhea, and gas
Vibriosis	*Vibrio parahaemolyticus*	Raw and cooked sea foods	Diarrhea, vomiting, headache and fever
Campylobacteriosis	*Campylobacter jejuni*	Poultry, raw meat, raw milk	Diarrhea, fever, abdominal pain
Listeriosis	*Listeria monocytogenes*	Milk, soft cheese, vegetables	flu-like illness mimics meningitis
E.coli food infection	*Escherichia coli*	Raw/ undercooked Foods- meat, cheeses	Profuse watery diarrhea, fever
Shigellosis	*Shigella sp*	Raw foods	Diarrhea, fever, vomiting
Yersiniosis	*Yersinia enterocolitica*	Under cooked pork, beef, and milk	Diarrhea, Pains mimicking appendicitis, fever
Hepatitis A	Hepatitis virus A,E	Raw/ undercooked shellfish, salads	Jaundice, fever, weakness, nausea
Viral gastroenteritis	Rota virus	Raw / mishandled food	Diarrhea in children
Amoebic dysentery	*Entamoeba histolytica*	Raw / mishandled food.	Dysentery, fever, chills

17.3. PREVENTION OF FOOD POISONING

1. The first step in preventing food poisoning is to assume that all foods may cause food-borne illness.
2. Good personal hygiene - washing hands, use of disposable aprons and gloves by processing staff prevents transmission of food pathogens.
3. Sanitizing food preparation surfaces, utensils and processing equipments thoroughly before and after handling raw foods prevents contamination and infection.

4. Avoid contaminated water, the major source of food pathogens.
5. Storage at temperature below the growth range of food pathogens would lessen the incidence of food poisoning.
6. Keep refrigerated foods below 5°C; serve hot foods immediately or hold them above 60°C, as the danger zone is between 5°C and 60°C.
7. Cook foods thoroughly and avoid eating under-cooked, under processed or under-pasteurized food; avoid eating raw sea foods.
8. Prevent cross contamination from raw material to cooked items to eliminate incidence of food poisoning.
9. Good manufacturing practices (GMP) should be followed in the manufacture, processing and storage of foods.
10. Reject suspected food.

QUESTIONS

1. Differentiate between food intoxication and food infection
2. Differentiate between exotoxin and endotoxin produced by food borne bacteria
3. Write an account on botulism, the causative agent, symptoms and toxigenic types.
4. List out various food pathogens, diseases caused and their symptoms.
5. Write an account on salmonellosis.
6. Comment on enteropathogenic *E.coli* strains.
7. What is Shigellosis?
8. Write on food-borne viruses and protozoa.
9. Write on prevention of food poisoning.
10. Botulinal toxin is a

    a) Mycotoxin b) Enterotoxin

    c) Endotoxin d) Exotoxin
11. Match the following

    1. Food infection - a. *Vibrio parahaemolyticus*
    2. Food intoxication - b. *Entamoeba histolytica*
    3. Mycotoxin - c. *Clostridium botulinum*
    4. Amoebiasis - d. *Clostridium perfringens*
    5. Sea foods infection - e. *Aspergillus flavus*

12. Which one of the following is the symptom of Salmonellosis?
    a) Thickness of speech
    b) Double vision
    c) Fatigue
    d) Nausea and vomitting
13. Which one of the following is the symptom of Botulism ?
    a) Thickness of speech
    b) Double vision
    c) Fatigue
    d) All the above
14. Which hepatitis cause food poisoning?
    a) Hepatitis A
    b) Hepatitis B
    c) Hepatitis C
    d) Hepatitis D
15. Botulin toxin E is found in
    a) Fruits
    b) Vegetables
    c) Marine foods
    d) Meat

18

MYCOTOXINS

Fungi cause diseases in plants and animals and in such diseases, the fungi actively grow and invade the body of their hosts. There is another means by which fungi can cause harm is growing on stored food material that we consume and produce toxic metabolites referred to as mycotoxins (means fungus poisons). Food is particularly susceptible to fungal infections at almost every stage of storage or production. Mycotoxin is toxic substance produced by a fungus. The term is restricted to toxins produced by fungi growing in human foods or animal feeds. The growth of the fungi in food grains leading to physical spoilage of grains and toxin production is a serious problem in developing countries. Their occurrence in food and feed has both economic and health implications, as the toxic effects in human beings, animals and other living systems were well established. The effects of poisoning by mycotoxin are referred to as mycotoxicoses.

18.1. NATURE OF MYCOTOXICOSES

Mycotoxicoses are generally more common in livestock than in humans and the diagnosis of these diseases are not easy. One must look for the following criteria:

1. The disease is not contagious and no microbe is found in the diseased tissue.
2. The disease can be correlated with the consumption of certain specific food/feed and regression is noted when the feed is withdrawn.
3. Examination of the food/feed reveals the presence of toxic fungi.
4. The fungi when isolated and cultures produce the toxin *in vitro*
5. Sometimes the toxin can be extracted from the food/feedstuff itself.

NATURE OF MYCOTOXIN

Mycotoxins are non-volatile, toxic secondary metabolites of low molecular weight produced by certain saprophytic and pathogenic fungi on food materials under both pre and post harvest conditions of tropical and subtropical regions. Mycotoxins are chemically unrelated group of fungal metabolites characterized by their ability to induce toxic responses in man and animals when food or feed containing them was consumed. Mycotoxins occur in great number and variety. Common toxins include alkaloids, cyclopeptides, and coumarins. The fungi produce mycotoxins, which are difurano- coumarins, in several forms and their effects are similar. Mycotoxins are chemically very stable, resistant against temperature, storage and processing conditions.

Human sufferings from mycotoxicoses include, ergot poisoning associated with ingestion of rye flour contaminated with ergot (holy fire/St. Anthony's fire), cardiac beriberi associated with *Penicillium* molds in rice (yellow rice toxin), and alimentary toxic aleukia (ATA), associated with *Fusarium* molds on over wintered millets.

Ergot is the common name of a fungus in the genus *Claviceps,* which is parasitic on certain grains and grasses. The fungus forms a *Sclerotium* in winter, and this small structure is what is usually referred to as 'ergot'. There are about 50 known species of *Claviceps,* most of them in the tropical regions. *C. purpurea* can affect a number of cereals including rye (its most common host), triticale, wheat and barley. The ergot sclerotium contains high concentrations (up to 2% of dry mass) of the alkaloid ergotamine. Ergot alkaloids have a wide range of biological activities including effects on circulation and neurotransmission.

Ergotism is the name for severe pathological syndromes affecting humans or animals that have ingested ergot alkaloid-containing plant material, such as ergot-contaminated grains. The common name for ergotism is "St. Anthony's Fire", in reference to the symptoms, such as severe burning sensations in the limbs. These are caused by effects of ergot alkaloids on the vascular system due to vasoconstriction of blood vessels, sometimes leading to gangrene and loss of limbs due to severely restricted blood circulation.

EFFECTS: The effects of ergotism can include hallucinations, skin inflammation, severe liver damage, hemorrhages, abortion, convulsions, neurological disturbances, and death in livestock and humans. Several mycotoxins, including aflatoxins have been linked to increased incidence of cancer in human beings. As the compounds are active at parts per billion, mycotoxins are among the most toxic and dangerous fungal metabolites. Mycotoxins also induce cancer of the liver. The toxins thus produced, when consumed unknowingly with the food or feed cause various diseases, acute or chronic and these diseases are collectively referred to as mycotoxicoses. Mycotoxins are neither contagious nor infectious, but can occur on a herd-wise basis. The effects in domestic animals include allergic reactions, reproductive failure, loss of appetite, feed refusal, decreased feed efficiency, suppression of immune system, and mortality.

ORGANISMS: Most of the mycotoxins are produced by cultures of the fungal genera like, *Aspergillus* , *Fusarium* , *Penicillium, Alternaria.* These include aflatoxins produced by *Aspergillus flavus* and *A. parasiticus,* trichothecene mycotoxin, deoxynivalenol (DON) produced by *Fusarium graminearum,* fumanosin produced by *Fusarium proliferatum* and Zearalenone produced by *Fusarium graminearum.* Molds such as *Myrothecium, Pithomyces, Stachybotrys, Trichothecium* and *Epicoccum* also produce mycotoxins. These fungi grow on stored grains and animal feeds when humidity is high enough. They also grow in living plants of cotton, peanuts and corn, where colonization of the host plant may take place prior to seed ripening. The fungi, and probably toxins, are present in the seed at harvest. Fungi produce a range of secondary metabolites, which include compounds that aid their competitive position. Among the compounds, a variety causes toxicosis in mammals, especially humans, insects and other herbivores. Spores of a wide range of fungi are common in the air. If conditions are suitable (high water activity and moderate temperature), fungi colonize the foodstuffs. Some common air-borne fungi are known to produce extremely toxic compound, these include *Aspergillus, Alternaria, Fusarium* and *Penicillium.*

18.2. AFLATOXINS

Aflatoxins are a complex of mycotoxins produced by species of the soil saprophyte *Aspergillus.* Aflatoxins are the first mycotoxins, discovered in 1960, produced by *Aspergillus flavus.* Aflatoxins were first reported in Brazilian groundnut meal in early 1960s which when fed caused death of thousands of turkey birds in England. Birds are particularly sensitive to mycotoxins and died from liver and kidney failure. Since then lot of work has been carried out on aflatoxin to establish their occurrence and to find out their physical, chemical and biological properties. The toxigenic strains of *Aspergillus flavus* and *Aspergillus parasiticus,* grow and produce aflatoxins on various agricultural commodities prior to and after harvest. Production of aflatoxin was not restricted to any geographical area or environmental conditions.

Mycotoxins occur in nuts (Fig. 18.1), cereals under conditions of high humidity and temperature and present a risk to human health that is insufficiently recognized. Of these mycotoxins, only aflatoxins have attracted the great attention all over the world. Aflatoxins are a closely related group of secondary metabolites produced by *Aspergillus flavus* and its related species in seeds and food products. The genus *Aspergillus* is a highly versatile genus growing in a variety of substrates and having wide ranges of tolerance. The morphological characteristics of *Aspergillus* spp are given in the Table 18.1. *Aspergillus flavus* constitutes a dominant air mycoflora. The two major *Aspergillus* species that produce aflatoxins are *Aspergillus flavus,* which produces only B mycotoxins, and *Aspergillus parasiticus,* which produces both B and G aflatoxins. Aflatoxin B_1 and G_1 are found most commonly. Aflatoxins are designated as B_1, B_2, G_1 and G2 in accordance with the blue and green colour fluorescence at differing Rf values under UV light (Fig. 18.2). Aflatoxins [M.sub.1] and [M.sub.2] are oxidative metabolic products of aflatoxins [B.sub.1] and [B.sub.2] produced by animals following ingestion of contaminated food, and so appear in milk (both animal and human), urine and faeces. Aflatoxicol is a reductive metabolite of aflatoxin.

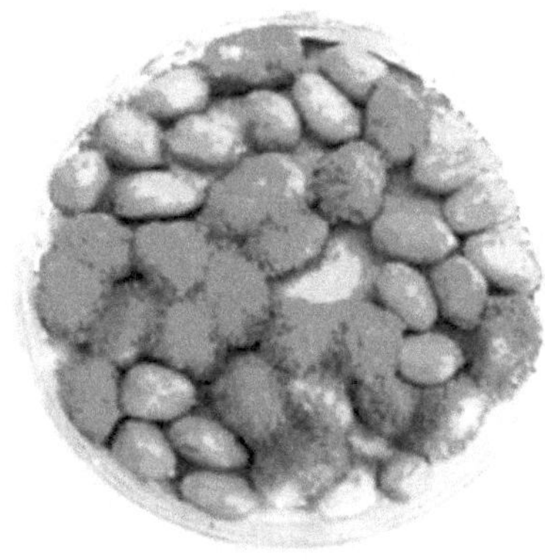

Fig. 8.1 Aflatoxin in ground nuts.

Aflatoxins are regarded as a serious health hazard. They may cause liver disease and may be carcinogenic. Aflatoxins are hepatocarcinogenic, mutagenic, teratogenic and immunosuppressive and have been classified as Clss I human carcinogens by the International Agency for Research on Cancer (IARC). It is heat stable and will withstand temperatures up to boiling. US, FDA has set an aflatoxin limit of 20 ppb for foods and feeds. European union has enacted aflatoxin tolerance level of 2 microgram per Kg for nuts and cereals for human consumption. Codex Alimentarius prescribes that the maximum level of aflatoxin M1 in liquid milk should not exceed 50 nanogram per kg.

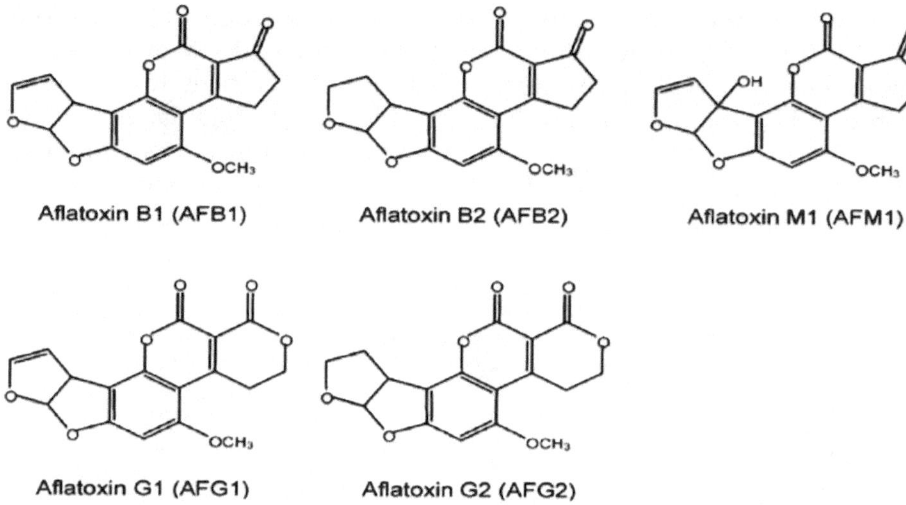

Fig. 18.2. Structure of Aflatoxins

TABLE 18.1. CHARACTERISTICS OF *ASPERGILLUS* SPECIES

CHARACTER	ASPERGILLUS FLAVUS	A.PARASITICUS	A. TERREUS
Colony diameter	4.0 to 4.5 cm	3.2 to 3.5	2.0 to 2.5
Obverse	Yellowish green, becoming green with age	Yellowish green turning dark green with age	Varying shades of cinnamom
Reverse	Greenish yellow	Greenish yellow	Yellowish brown
Head	Radiating, becoming loosely columnar with age	Radiate	columnar
Stipe	Long verrucose, hyaline	Long and rough	Short smooth
Vesicle	Dome shaped, fertile on entire surface	Globose, fertile on entire surface	Hemisphericle, fertile in upper 1/3 area
Metulae	Present small	Few or absent	Present almost parallel to stipe axis
Phialides	Small. Ampulliform	Ampulliform with long broad neck	Ampulliform, thin
Conidia	globose to subglobose, usually rough, yellowish green	Globose, very rough often showing connective tissue, greenish	Globose in long chains compact column of uniform diameter
Mycotoxin	B1, B2, Aspergillic acid, cyclopiazonic acid, dihydroxyaflavinine	Aflatoxin B1, B2, G1,G2,aspergillic acid, kojic acid	Citreoviridin, citrinin, gliotoxin, patulin, terrain, terreic acid

The chemical composition of seed also influences mycotoxin production. Although the pattern of colonization of different substrates is similar, there may be specific differences for products rich in oil, protein or starch. For example, *Aspergillus flavus* is more likely to occur and present a greater toxin hazard in oil seeds than in other grains. *Fusarium*, a field fungus is also a potent toxin producer, especially in maize. Conidia in *Aspergillus* spp in (Fig. 18.3).

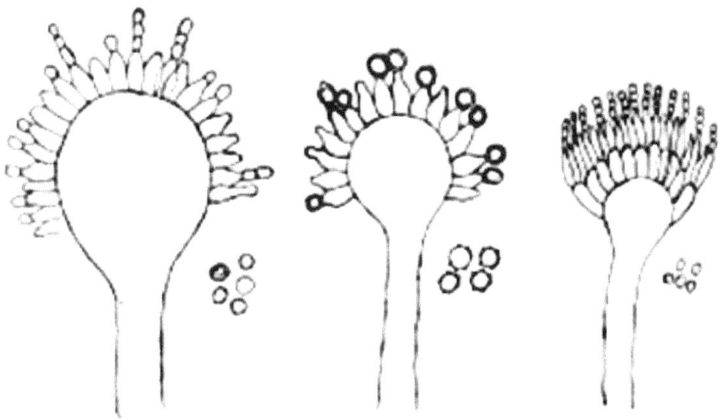

Fig. 18.3. *Aspergillus flavus, A.parasiticus, A. terreus*

OCHRATOXINS

Ochratoxins are secondary metabolites of *Aspergillus* and *Penicillium* strains, found on cereals, coffee and bread, as well as on all kinds of food commodities of animal origin in many countries. The most frequent is ochratoxin A, which is also the most toxic. It has been shown to be nephrotoxic, immunosuppressive, carcinogenic and teratogenic in all experimental animals tested so far.

TRICHOTHECENES

Trichothecenes are mycotoxins produced mostly by *Fusarium graminearum*, although other genera, *Trichoderma, Trichothecium, Myrothecium* and *Stachybotrys* are also known to produce these compounds. To date, 148 trichothecenes have been isolated, but only a few have been found to contaminate food and feed. The most frequent contaminants are deoxynivalenol (DON), also known as vomitoxin, nivalenol (NIV), diacetoxyscirpenol (DAS), while T-2 toxin is rare. Vomitoxin causes pigs and other animals to refuse food containing this mycotoxin.

FUMONISINS

Fumonisins are mycotoxins produced throughout the world by *Fusarium moniliforme* and related species when they grow in maize. Fumonisins [B.sub.1] and [B.sub.2] are of toxicological significance, while the others (B.sub.3], [B.sub.4], [A.sub.1] and [A.sub.2]) occur in very low concentrations and are less toxic. Fusarium toxins are apparently not carried over into milk, meat and eggs.

FUMAGILLINS

Compound produced by *Aspergillus fumigatus*, although classified as a mycotoxin, it is also used effectively as an antibiotic for amoebic parasites. However dosage is very important, if too much is used it can be fatal.

ZEARALENONE (ZEA)

Zearalenone, is a potent estrogenic metabolite produced by some *Fusarium* and *Gibberella* species. Particularly, is produced by *Fusarium graminearum, Fusarium culmorum, Fusarium cerealis, Fusarium equiseti, Fusarium verticillioides,* and *Fusarium incarnatum*. Zearalenone is heat-stable and is found worldwide in a number of cereal crops, such as maize, barley, oats, wheat, rice, and sorghum.

18.3. FOOD SAFETY

The enumeration and identification of toxic fungal species in grain and food materials are essential factors for quality assurance and control for both raw materials and finished products. The early detection of aflatoxigenic fungal species serves as a warning signal of potential health hazard. The growing awareness towards procuring food free from microbial hazards, focus on the development of improved methods that are rapid and capable of detecting very low, up to ppt levels of toxins were made. Biotechnological methods such as enzyme linked immunosorbent assay (ELISA), radioimmuno assay (RIA), immuno affinity column methods are being used for specific detection and estimation of mycotoxins. These methods use specific antibodies raised against the toxin and are very specific and sensitive for the detection.

Control Measures: Key management steps to minimize grain mold and mycotoxin contamination.

1. Ensure proper storage conditions - grain moisture, temperature and relative humidity.

2. Minimize mechanical damage - harvest and post-harvest shipping and handling.
3. Minimize insect pest damage - pre-harvest and post-harvest storage.
4. Sanitation of storage facility - critical management practices.
5. Mycotoxin tolerant hybrids - some tolerant hybrids available.
6. Detoxification

1. Requirement for Proper Storage Conditions

To minimize the potential for mold growth and mycotoxin production, grain moisture content should be brought down to safe level, for instance 13% of maize or 8% of ground nut. The moisture content of grains should be reduced to less than 15 percent within 48 hours after the grain is harvested. Field drying and mechanical drying were found most effective in controlling aflatoxin contamination in grains. This may be difficult to achieve in many storage facilities, especially larger capacity elevators. A storage temperature of less than 40° C should be maintained. Good air circulation throughout the storage bin is important in order to prevent the formation of localized pockets of high moisture in the grain mass that would allow the fungal growth. Storage conditions optimal for maintaining grain quality will minimize mold development and mycotoxin contamination. Modified atmosphere containing 60% carbon dioxide and 20% oxygen were more effective in controlling mold growth and aflatoxin production. Biogenerators for carbon dioxide production could be used to store grains.

2. Minimizing Mechanical Damage

Harvest and postharvest grain handling should be designed to minimize mechanical damage. Although most mold pathogens can directly penetrate plant tissues, mechanical damage provides additional entry sites, facilitating infection and spread from kernel to kernel during shipping and storage.

3. Minimizing Insect Damage

Like mechanical damage, insect feeding provides additional entry sites for mold pathogens, facilitating infection and spread from kernel to kernel during shipping and storage. Some insects also act as vectors of grain mold pathogens. An effective integrated insect management plan during the growing season as well as during postharvest, shipping and storage of the grain can reduce mold and mycotoxin contamination.

4. Sanitation of Storage Facility

The grain mold pathogens are well adapted to survive for long periods within grain storage facilities. Storage bins should be cleaned thoroughly each season. The type and configuration of the storage facility will dictate the nature of the sanitation required. The objective is to remove old grain and residue that could provide a harborage for these fungi. Fungicides are not currently used to manage grain molds or mycotoxin contamination. There are no fungicidal fumigants labeled for use in grain storage facilities. Antifungal agents such as propionic acid and sorbic acid are used in some storage facilities with varied success in reducing grain mold and mycotoxin contamination.

5. Mycotoxin Tolerant Hybrids

The ideal remedy would be the breeding of varieties resistant to attack by mycotoxigenic fungi. Conventional and molecular approaches to plant breeding may yield mycotoxin resistant crop varieties. Such varietal resistance has been reported in maize, sorghum and ground nut. The wild or native varieties of grain crops that possess resistance to mycotoxigenic fungi can be crossed with ruling high yielding varieties to develop mycotoxin tolerant hybrids.

6. Detoxification

In the absence of suitable methods for prevention of toxin formation, the next best remedy is to evolve methods of detoxification. The physical methods recommended are solvent extraction, adsorption on colloidal materials like bentonite clay and aluminosilicates, inactivation by heat, irradiation with UV or gamma rays. Treatment with 4 kGy irradiation inhibited aflatoxin production. Chemical method using oxidizing agents such as ozone or hydrogen peroxide was found effective. Detoxifying chemicals, ammonia, sodium hypochlorite, ammonium bis propionate, and propionic acid are widely recommended. Extracts of medicinal plants have been reported to inhibit aflatoxin synthesis. The biological control of *Aspergillus flavus* was also tried. Microorganisms, *Flavobacterium auranticum, Pitchia guilliermondi* and *Lactobacillus* spp have been found to degrade aflatoxins. *Saccharomyces cerevisiae* was effective in reducing fumonisin production in grains.

QUESTIONS

1. Define

   i) Mycotoxin ii) Mycotoxicosis

   iii) Aflatoxin iv) Fumonisin

2. What are the effects of mycotoxins in domestic animals and human beings?
3. List out fungi producing mycotoxins and their nature
4. Comment on ergotism.
5. Write a brief account on aflatoxins.
6. What are the characteristics of aspergillus species?
7. Explain the food safety aspects of mycotoxins.
8. Describe various management steps to be followed to prevent mycotoxin contamination in foods.
9. Aflatoxins are a series of toxins typically accumulate in

   a) Eggs b) Fish

   c) Peanuts d) Dairy products

10. Aflatoxin B_1 and B_2 are produced by

    a) *Aspergillus flavus* b) *Aspergillus parasiticus*

    c) Both a & b d) All the above

11. Aflatoxin B and G are produced by

    a) *Aspergillus flavus* b) *Aspergillus parasiticus*

    c) *Aspergillus niger* d) All the above

12. Aflatoxins are transmitted through

    a) Food grains b) Fish

    c) Canned food d) None of the above

13. Which of the following are mycotoxins?

    a) Ochratoxins b) Fumonisins

    c) Trichothecenes d) All the above

14. The chemical used for de-toxification of mycotoxins is

    a) Mercuric chloride b) Phenol

    c) Sodium hypochlorite d) Ethylene

19

GENETICALLY MODIFIED FOODS

GENETIC ENGINEERING

Genetic engineering is a means of improving the food supply even before harvest or slaughter by improving yields, increasing disease resistance, and enhancing the nutritional qualities of various foods. Broadly speaking, genetic engineering refers to any deliberate alteration of an organism's DNA. Genetic engineering has been practiced for thousands of years, ever since humans began selectively breeding plants and animals to create more nutritious, better tasting foods. In the past two decades, genetic engineering has become increasingly powerful as scientific advances have enabled the direct alteration of genetic material. Genetic engineering is alternately called as ' recombinant DNA technology' or 'gene cloning'. Genes have been cut and pasted from one species to another, yielding, for example, disease-resistant squash and rice, frost-resistant potatoes and strawberries, and tomatoes that ripen-and therefore spoil-more slowly.

Genetically modified organism (GMO) is a microbe or more complex plant or animal in which DNA has been altered for a particular purpose such as manufacturing proteins, making improvement to the organism, research into the nature of genes and biological process, correcting genetic defects.

ADVANTAGES OF GENETICALLY MODIFIED FOODS

1. It is faster and lower in cost to develop GM crops/ animals compared to conventional breeding techniques. Desired changes can be achieved in very few generations.
2. It allows greater precision in selecting characteristics. It reduces the risk of random occurrence of undesirable traits.
3. It makes it possible to cross the species barrier.

4. GM crops resistant to abiotic stress such as cold, drought, flood, salt, heat have been developed.
5. Pest and disease resistant, crops reduced reliance on chemical pesticides.
6. GM crops supply alternate resources to industries in the form of pharmaceuticals, starches, cocoa butter.
7. GM crops helped to reduce post harvest losses of foods.
8. GM crops enhanced nutritional value of food.
9. GM crops increased efficiency of mineral usage and maintain soil fertility.
10. Transgenic animals produced bioactive products, like human alpha lactalbumin.

19.1. APPLICATION OF GENETIC ENGINEERING IN CROP IMPROVEMENT

1. Pest resistance

The genes encoding for Bt (*Bacillus thuringiensis)* crystal toxin are carried in plasmids and transferred to crop plants to make them resistant to pests. Plants carrying Bt genes can produce Bt protein, the toxin which interferes with ion transport and feeding of insects, boll worms. These transgenic insecticidal crops promise to revolutionize Integrated pest management (IPM), for many crops (Bt cotton, tomato, corn, potato, cabbage, soybean, tobacco, sweet corn, brinjal). Bt crops give effective protection against targeted pests, less use of pesticides and improved crop yield and quality. Transgenic plants resistant to nematode (*Meloidogyne incognita)* were developed by silencing m-RNA.

2. Disease resistance

Disease resistance in crops is obtained by treating the plant cells with toxins of plant pathogens. Potato varieties resistant to late blight were developed by exposing the cells to *Phytophthora infestans* toxins. The toxin resistant callous regenerates were resistant to the disease. Genes for antifungal, bacterial and viral compounds have been transferred into crops to protect them from diseases.

3. Herbicide resistance

The herbicide detoxifying genes present in bacteria have been transferred to plants. These genes could confer resistance to widely used herbicides like 2,4-D, glycophosphate, bromoxynil. These new crops provide farmers with greater flexibility in herbicide use and reduce the need for pre-emergent soil applications in many crops. Round up Ready soybean, cotton, corn, canola and sugar beet are transgenic herbicide resistant crops.

4. Adaptation to Harsh Growing Conditions

This includes conditions such as drought, temperature extremes, and soils with high salt contents. Potato with an antifreeze gene from cold water fish is able to tolerate cold temperatures that normally would kill unmodified seedlings. Modifying linoleic acid production improves survival at low temperature. The genes associated with water use efficiency and osmoregulation (OSM) were identified in plants (Salvinia, *Arabidopsis*) and bacteria (halophiles). Isolation and transfer of these genes confer drought and salt tolerance to crop plants.

5. Nutritional Value of Crops

The genes encoding major seed storage proteins are available. The alteration of their coding sequence so as to increase the level of certain amino acids and their introduction is possible - tryptophan in maize, lysine in rice, threonine in wheat, sulfur amino acids in legumes. i) Oil quality is improved in canola and soyabeans by modifying the fatty acid content with high lauric acid and high oleic acid respectively. ii) A rice variety, Golden rice, containing β-carotene and iron has been developed iii) Introducing gluten gene from bacteria in to wheat improved the quality of bread.

6. Improved Quality of Foods

Pectin is an essential component of many fruits and it is naturally broken down by polygalacturonase enzymes. Anti-sense technology can be used to neutralize the production of polygalacturonase, delaying the softening of fruits once they have been harvested. Genetic modification offers the opportunity to delay the onset of ripening by suppressing poly galacturonases, the cell wall degrading enzymes. This allows the fruit to stay on vine longer and develop a greater flavour and colour e.g. Calgene's Flavr Savr tomato (Fig. 19.1). Genetic modification could decrease or eliminate the protein allergens that occur naturally in specific foods. By transfer of antifreeze genes from fish to tomato, strawberry , long cold storage of fruits and vegetables is achieved.

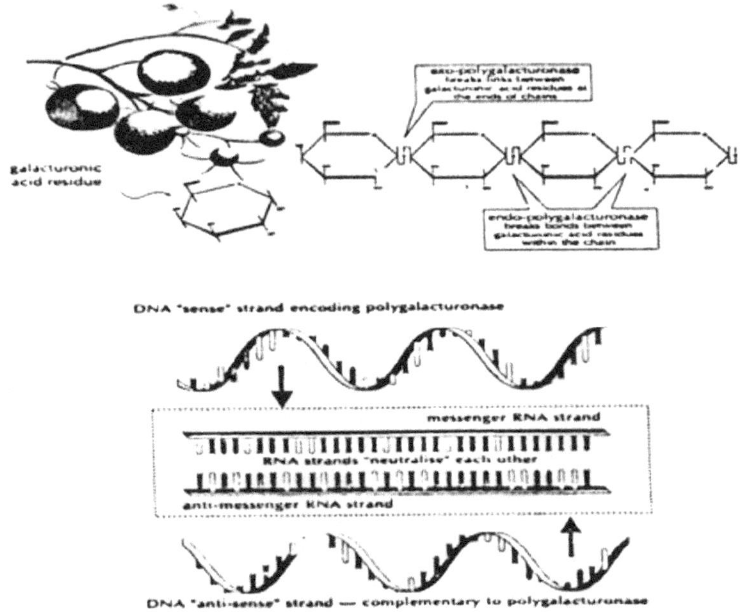

Fig. 19.1. Slow ripening tomato by antisense RNA technology

7. Reduction of Antinutritional Factors

Anti-nutritional factors are substances in any plant products which render them unsuitable or less suitable as food. Khesari dal (*Lathyrus sativus*) contains a toxic amino acid, which causes paralysis; Gossypol in cottonseed decreases food intake in animals. Lima beans contain antitrypsin that inactivates enzyme trypsin. Genetic modification offers the opportunity to decrease or eliminate these anti-nutritional factors naturally present in specific foods. Antisense RNA technique prevents the formation of gene product, combined with plant breeding process, have led to production of cultivars with low concentration of toxic substances.

8. Pharmaceuticals

Medicines and vaccines often are costly to produce and sometimes require special storage conditions not readily available in third world countries. Researchers are working to develop edible vaccines in tomatoes and potatoes. These vaccines would be much easier to transport, store and administer than traditional vaccines. Production of value added products like nutrition supplements, pharmaceuticals, starches, cocoa butter using transgenic crops is known as 'molecular farming'.

Genetic engineering represents the ultimate tool as it extends humanity's reach over the forces of nature as no other technology has done ever before. Transgenic plants engineered for resistance to pests, disease and herbicides and for nutritional quality are now being produced.

GOLDEN RICE

Rice, the staple food to 3.5 billion Asian people, lack pro vitamin A. Vitamin A deficiency leads to blindness and weakens immune system in children. Dr. Ingo Potrykus, Genetist Swiss Federal Institute of Technology and Dr. Peter Beyer, Biochemist, University of Freiburg, Switzerland developed golden rice, genetically engineered rice to produce β-carotene, pro vitamin A.

METHODOLOGY (Fig. 19.2)

- β-carotene genes were isolated from Daffodil plant, *Erwinia* bacteria.
- The genes where inserted into the plasmid vector - Ti Plasmid of *Agrobacterium tumefaciens* in T-DNA region.
- These agrobacteria then added to a petri dish containing rice embryos. As they infect they also transfer the genes that encode for making beta carotene. Embryos grown in to rice plants.
- Provitamin producing transgenic rice plant selected. The transgenic rice plants must be crossed with strains of rice that are grown locally and are suited to a particular region's climate and growing conditions.

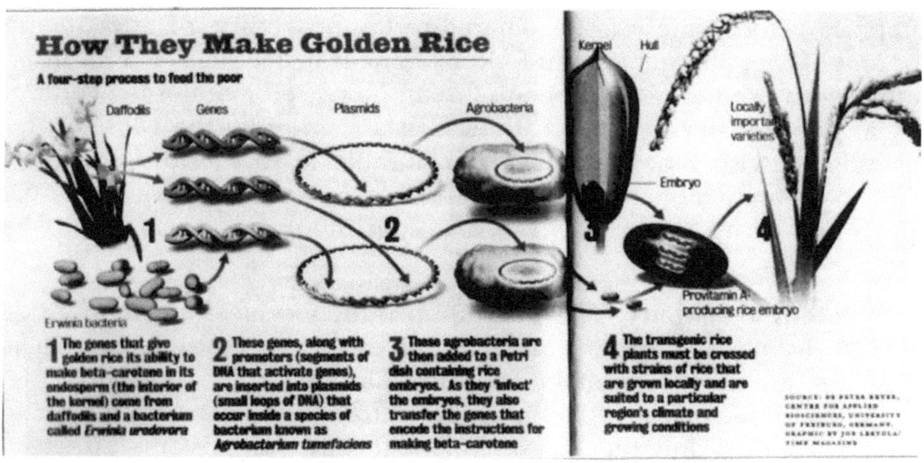

Fig. 19.2. Producing Golden rice

19.2. TRANSGENIC ANIMALS

A transgenic animal contains in its genome, a gene or genes introduced by transfection. The sheep, "Dolly" is the first transgenic animal. Transgenic animals produce biological products, such as antitrypsin protein used to treat 'emphysema. In 1997, the first transgenic cow, **Rosie** produced human protein enriched milk. The milk contained the human alpha lactalbumin and nutritionally more balanced product for human babies than natural cow milk. Genetically modified salmon fish was the first transgenic animal obtained for food production. Rabbits are quite promising for gene farming , which aims at production of recoverable biologically important proteins encoded by transgenes. Transgenic sheep have been produced to achieve better growth and meat production.

19.3. FOOD SAFETY CONCERNS EXPRESSED ABOUT GM FOODS

Antibiotic resistance: Antibiotic resistant genes are used routinely in both plant and microbial genetic engineering as selectable markers. When the gene of interest is closely linked to an antibiotic resistant gene, it is relatively easy to detect those cells which have been successfully transformed. Thus, many of the current generation genetically modified crops contain these genes. A major concern is whether the transfer of antibiotic resistance from a marker gene in plant to a microorganism naturally present in the human gut or pathogens is possible. Alternative markers such as fluorescent genes may be employed.

Allergenicity: The presence of allergens in food is already an important food safety issue. Therefore the possibility of introducing or amplifying allergenicity of food by gene transfer is a concern. In an attempt to produce a soya bean with an increased methionine content by a gene transfer from brazil nut, it was found that the allergenicity associated with Brazil nut was transferable to soya beans. To ensure GM foods are allergen-free, a careful assessment procedure for novel proteins, using a combination of experimental testing should be employed.

Potential toxicity: The possible production of toxic compounds in GM foods is another concern. An example of this scenario is demonstrated by the research conducted in Rowett Research institute, which involved inserting lectins into potatoes, as a means of enhancing pest resistance. But they were shown to have an adverse effect on rat during feeding trials. The experiments were poorly designed and demonstrated the need for greater awareness in assessing the safety of GM foods.

Environmental issues: The most wide spread concern is the effect of GM crops on the environment. There are fears of genetic drift, for example pollen from one kind of plant is taken up by another plant - cross contamination. Similarly, genetically engineered triats, such as herbicide resistance, may spread to weeds leading to the creation of indestructible weed immune to herbicides (super weed). It has also been suggested that the adoption of insect resistant crops will disrupt the food chain and lead to the extinction of insect species, thereby reducing bio-diversity.

Socio-economic concern: Bringing GM food to market is a lengthy and costly process, and of course Agri-Biotech companies wish to ensure a profitable return on their investment. Patenting these new plant varieties will raise the price of seeds so high that small farmers will not be able to afford seeds for GM crops. To combat possible patent infringement they may introduce suicide gene (terminator technology) into GM crops. They would be viable for only one season and would produce sterile seeds that do not germinate. Farmers would need to buy a fresh supply of seeds each year.

Ethical concern: It is important to pay attention in particular to ethical concerns which some groups have about particular foods and consider they may be affected by use of genetic modification. Some groups may object on ethical grounds to the consumption of food containing copy genes from humans or animals, particularly animals which are the subject of religious dietary restrictions. These genes are referred as ethically sensitive. For example, Muslims, Sikhs and Hindus have ethical objections to consuming foods containing copy genes from certain animals such as pigs and cows. Similarly, strict vegetarians would object to incorporation of copy genes from any animal into plant food.

19.4. CONCLUSIONS

- GM foods have potential to solve hunger and malnutrition problems.
- GM technology has the potential to help, protect and preserve environment by increasing the yield and reducing reliance upon chemical pesticides and herbicides.
- This technology requires careful regulation. The minimum requirement should be that the GM food should be as safe as the original product and present no threat to human health.

- To ensure public confidence on safety of GM foods, adequate labeling and provision of accurate product information are essential.
- Governments, scientists and industries should address the preferences and requirements of the consumer. Genetic engineering Approval committee, Government of India will make decisions on safety of GM organisms (plants, animals, microorganisms) for public service.

QUESTIONS

1. What is genetic engineering?
2. What are the advantages of genetically modified foods?
3. Discuss the applications of genetic engineering in crop improvement
4. Write on improvement of nutritional value foods through genetic engineering.
5. Comment on golden rice
6. Discuss the applications of transgenic animals.
7. Discuss food safety concerns expressed about GM foods
8. Golden rice is rich in this vitamin

   a) Biotin b) Vitamin D

   c) Vitamin A d) Riboflavin
9. The transgenic animal Rosie is a

   a) Buffalo b) Sheep

   c) Cow d) Rabbit
10. The enzyme involved in slow ripening of transgenic tomato

    a) Amylase b) Lipase

    c) Polygalacturonase d) Glucoisomerase

20

EXPERIMENTS IN FOOD AND DAIRY MICROBIOLOGY

Ex.No.1. MICROBIAL SPOILAGE OF FRUITS

Fruits are living structures that continue to perform metabolic reactions and sustain physiological processes for a considerable length of time after harvest. The post-harvest infection of fruits by microorganisms may cause serious losses of fruits both in terms of quantity and quality. Initial infection which may be the direct result of mechanical or insect damage is followed by invasion of fruit tissues by secondary infection by common saprophytic organisms. . Many bacteria and fungi, for instance, are involved in decay after harvest. Typical fungi include *Alternaria, Botrytis, Colletotrichum, Fusarium, Monilia, Penicillium,* and *Rhizopus*. These fungi are generally weak pathogens, in that they usually infest only weak or damaged fruit. Molds are the most predominant spoilage organisms in fruits. Fungal rots in fruits like apple show brown or cream coloured areas.

Bacterial soft rot caused by *Erwinia carotovora* is wide spread in fruits that are not very acid. Fruits like straw berries, melons are in direct contact with moist soil and become infected from that source.

MICROBIAL SPOILAGE OF FRUITS

FRUIT	SPOILAGE	ORGANISM
MANGO	Black mold rot, Anthracnose	*Aspergillus niger, Colletotrichum sp.*
PINEAPPLE	Chocolate rot, Soft rot	*Sclerotium rolfeit,* yeast injury
PAPPAYA	Soft rot, Anthracnose	*Botryodiplodia theobromae Colletotrichum gloeosporoides*
BANANA	Anthracnose, Black mold rot	*Colletotrichum musae, Aspergillus niger*

Procedure : The spoiled portions from fruits are cut with sterile knife and the microbial growth in the spoiled area is scrapped on to a glass slide with a drop of water. The cells, spores and hyphal structures are observed under microscope and identified based on their morphological characteristics. The organisms are stained and mounted as permanent slides.

Observations :......................

Ex.No.2. MICROBIAL SPOILAGE OF VEGETABLES

The harvested edible portions of plants, vegetables are exposed to microbial spoilage depending on several factors such as cell wall thickness, moisture content, inherent chemical substances etc. Vegetables are moisture rich neutral pH food, relatively more prone to microbial attack. The location of plant part used as vegetable is an important factor in microbial contamination and spoilage. The underground parts like roots, tubers or bulbs as in radish, beet, carrot, potato, onion are in direct contact with moist soil and become form the source. Vegetables like cucumbers, tomatoes may be in direct contact with the surface soil. Leaves, stems and flowers as in lettuce, greens, cabbage, asparagus, and cauliflower are generally infected by plant pathogens from the field.

SPOILAGE	ORGANISM	VEGETABLES
BLACK ROT	*Xanthomonas campestris*	Cabbage, cauliflower
BACTERIAL SOFT ROT	*Erwinia carotovora*	Carrot, cabbage, green beans
GRAY MOLD ROT	*Botrytis cinerae*	Carrot, cabbage, green beans
BLACK ROT	*Alternaria sp*	Carrot, cabbage, green beans
BLUE MOLD ROT	*Penicillium sp*	Onions
ANTHRACNOSE	*Colletotrichum sp*	Green beans, tomato, onion
FUSARIUM ROT	*Fusarium sp*	Carrot, potato
RHIZOPUS SOFT ROT	*Rhizopus stolonifie*r	potato, cabbage, radish

Fungal spoilage of vegetables often results in water soaked mushy areas. The fungal mycelia growing in the tissue below the skin and the aerial hyphae and spores may appear later. Molds are the most predominant spoilage organisms in vegetables.

Procedure : The spoiled portions from vegetables are cut with sterile knife and the microbial growth in the spoiled area is scrapped on to a glass slide with a drop of water. The cells, spores and hyphal structures are observed under microscope and identified based on their morphological characteristics. The organisms are stained and mounted as permanent slides.

Observations : ....

Ex.No.3. SPOILAGE OF BAKERY PRODUCTS

Bread is contaminated by mold spores in air during cooling and wrapping. The moisture content of commercially produced bread is insufficient for the growth of most organisms except molds. Moldy growth in bread is generally produced by, Bread molds- *Rhizopus stolonifer* and *Mucor* sp and also *Aspergillus niger* and *Penicillium expansum*. Ropiness in bread is caused by *Bacillus subtilis*. Red bread is due to pigmented bacteria- *Serratia marcescens*. The baking process is sufficient to destroy microbes in cakes, but icings, toppings, nuts and fruits added after baking can be the source of molds in cakes.

Materials required: Spoiled bread, Czapek's agar, pertiplates, pipettes, water blanks, and forceps.

Procedure : The samples are weighed and powdered and diluted in sterile water blanks. Plate out dilutions 10^{-3} and 10^{-4} using Czapek's agar medium, Incubate the plates for five days at room temperature and count the colonies.

Observations : ....

Ex.No.4. MICROBIAL SPOILAGE OF SEA FOODS

The bacteria most often involved in the spoilage of sea foods are part of the natural flora of the external slime and intestinal contents. The spoilage bacteria of the genera *Micrococcus, Bacillus, Escherichia, Proteus, Serratia, Sarcina* and *Clostridium* would mostly grow at ambient atmospheric temperature, whereas *Pseudomonas, Flavobacterium, Achromobacter* are found in chilling temperature.

Materials required: Fresh fish, salted fish, water blanks, pipettes, petri plates, nutrient agar medium with 3.5 % sodium chloride.

Procedure : The samples are weighed and thoroughly washed and diluted in sterile water blanks. From the serial dilutions plate out dilutions 10^{-3} and 10^{-4} for salted fish and 10^{-4} and 10^{-5} dilutions for fresh fish using nutrient agar medium with 3.5% Na Cl. Incubate the plates for two days at room temperature and count the colonies.

Result: ....

Ex.No.5. ISOLATION OF MYCOTOXIGENIC FUNGI FROM FOOD GRAINS

Mycotoxins are chemically unrelated group of fungal metabolites characterized by their ability to induce toxic responses in man and animals when food or feed containing them was consumed. Mycotoxins are the toxic secondary metabolites produced by certain saprophytic and pathogenic fungi on food materials under both pre and post harvest conditions of tropical and subtropical regions. Their occurrence in food and feed has both economic and health implications. Of these mycotoxins, only aflatoxins have attracted the great attention all over the world. As the storage fungi can withstand very high osmotic pressure, a high-osmotic sucrose medium is used as a selective medium for screening the food grains for isolation of mycotoxigenic fungi.

Material required: Petri plates, seeds, filer paper, surface sterilizing chemicals and Sucrose medium

Composition of sucrose medium (Christensen, 1957)

Sucrose	-	500 g
Sodium nitrate	-	3 g
Dipotassium hydrogen phosphate	-	1 g
Magnesium sulphate	-	0.5 g
Potassium chloride	-	0.5 g
Ferrous sulphate	-	0.01 g
Agar	-	20 g
Distilled water	-	1000 ml
pH	-	6.5

PROCEDURE: The various grains *viz.*, maize, sorghum, cumbu, ragi, soybean, groundnut and gingelly both unsterilized and surface sterilized with 0.1 % mercuric chloride solution for 60 seconds and washed repeatedly with 7.5 per cent sodium chloride, are placed in the sucrose medium by direct plating technique using sterile forceps. A marked paper template i.e. blue print is kept under the plates for even distribution of seeds. The number of grains plated from the samples *viz.*, maize, sorghum, cumbu, ragi, soybean, groundnut and gingelly are 10, 25, 25, 45, 10, 10 and 30 respectively. The number of grains placed in the plate depended on the size of the grain. The plates are incubated at room temperature at 30°C and observed from 3rd to 5th day. The infection per cent of unsterilized and surface sterilized seeds were determined using the following formula,

$$\text{Infection per cent} = \frac{\text{Number of seeds infected}}{\text{Total number of seeds plated}} \times 100$$

The isolated fungi are observed under microscope and identified based on their morphological characters.

Ex No.6. DIRECT MICROSCOPIC METHOD OF COUNTING BACTERIA IN LIQUID FOODS

The number of bacteria in liquid foods such as juice, milk can be determined by direct microscopic method. In this method a measured volume of liquid food is spread over a measured area of a glass slide in a thin film, which is dried, fixed, stained and observed microscopically. Since the area of a single microscopic field, number of microscopic fields per unit area and the volume of food examined are known, the average number of bacteria in microscopic field can be counted and the total number of bacteria per ml of liquid food is computed.

Materials required: Liquid food samples, pipette, glass slides, marker pen, methylene blue 0.6 % solution, and microscope with oil immersion objective.

Procedure: In clean glass slide one square cm area is marked in the centre of the slide, with a marker pen. By using a graduated pipette 0.01 ml liquid food sample is placed in the one cm area marked in the slide. Using a clean needle spread the liquid food sample uniformly in the marked area. Air dry the film slowly on a perfect level surface. Stain the smear with 0.6 % Methylene blue solution for two minutes. Drain the excess stain and dry the slide. Wash the slide with distilled water, drain and air dry. Using oil immersion objective of compound microscope determine the number of bacteria per microscopic field.

The area of the microscopic field is determined from the diameter of the field in eye piece, using the formula $\pi d^2/4$

1. Area of the microscopic field : $\pi d^2/4$
2. No. of microscopic fields per sq.cm, n : 10,000 x 10,000/ area of mic. field
3. Average No. of bacteria per microscopic field : X
4. No. of bacteria per ml : Average No. of bacteria x No. of microscopic fields per sq.cm x dilution factor : X x n x 100

Result: ....

Ex.No.7. EXAMINATION OF UNSPOILED CANNED FOOD FOR STERILITY AND QUALITY

Introduction

The credit to discovering the art of canning goes to Nicholas Appert (1810) who is known as father of canning industry. He believed that preservation of canned foods depended on the exclusion of outside air. It was Louis Pasteur (1860) who proved that microbes are the real cause of spoilage. The general outline followed for microbiological examination of cans was based on the recommendations laid down by Tanner (1944).

Material required: Canned food, sterile water, pipette, Petri plates, media, and tin opener.

Procedure: The particulars regarding the product are noted in the prescribed proforma given below. The can is thoroughly washed with hot soapy water or detergent solution, rinsed and then immersed in a solution of bactericide for 5-10 minutes. The can is opened with the help of a tin opener.

Transfer 10 ml of the liquid (juice) into 90 ml water blank; allow it to stand for 10 minutes. From this sample plate out 10^{-1}, 10^{-2} dilution with potato dextrose agar and 10^{-2} and 10^{-3} dilutions with nutrient agar. Incubate the plates at room temperature for five days then make the counts and differentiate types of organisms by colony morphology.

Particulars of canned food

1. Name of the product
2. Name of the manufacturer
3. Date of manufacture and age of the product
4. Size of the container (Dimensions) in cm
5. External appearance
6. Internal pressure
7. Head space (cm)
8. Gross weight (g)
9. Net weight (g)
10. Drained weight(g)
11. Internal appearance
12. Solid: appearance, colour, texture

13. Liquid: appearance, colour, ,taste and smell
14. Total soluble solids
15. p^H
16. Remarks (FPO- Fruit product order number)

Result: ....

Ex.No.8. ISOLATION OF LACTIC ACID BACTERIA FROM CURD

CURD (Dahi) is prepared by souring milk with a mixed culture of lactic acid bacteria, the starter. The milk is boiled and cooled. The starter culture is added and incubated for 12 hrs at room temperature. When the culture develops 0.80 to 0.90 % lactic acid, the fermentation is stopped by cooling. The curd is broken by agitation to enhance the flavour and odour of cultured butter milk. The starter culture for curd fermentation generally contain *Streptococcus lactis, S. cremoris, Leuconostoc citrovorum,* and *L. dextranicum*

Materials required : curd, water blanks, perti plates, pipettes and

Yeast extract lactose agar Composition :

Peptone	1.0 g
Yeast extract	0.5 g
K_2HPO_4	0.2 g
Lactose	0.5 g
Bromocresol purple	0.2 ml
(1.6 g in 100 ml)	
Distilled water	100 ml

PROCEDURE : The curd sample is serially diluted up to 10^{5} dilution. Plated one ml each from dilutions 10^{-4} and $10^{-5.}$ The plates are incubated at room temperature for two days. After the incubation period the colonies are counted and population of lactic acid bacteria per ml of curd is estimated.

Result : ....

Ex.No.9. ISOLATION AND ENUMERATION OF YEAST FROM RICE BATTER

Introduction: Fermentation is one of the oldest method of food preservation. Fermented legume and cereal products form an important part of human diet in India. These fermented foods are also becoming popular in Western countries due to their high nutritive value and organoleptic characteristics. Some of the most common indigenous fermented food are: idli, dosa, bhallae, bhatura, jelabie, nan, pappadam, vadai and warri.

Idli: Rice and black gram mixtures are fermented in the proportion of 3:1 the ingredients are washed and soaked separately. Then ground, mixed and finally allowed to ferment overnight - ripening time from 16 hours at 31°C to 10 hours at 41°C. Due to microbial activity the batter is acidified and leavened. During fermentation the volume of idli batter increases ranging from 1.6 to 3.1 times its original volume. The p^H decreases from 6.0 to 4.3and the acidity increases.

Both bacteria and yeasts have been reported to participate in Idli fermentation and are generally introduced by the two ingredients. Black gram provides lactic acid bacteria inoculums and serve as substrate for fermentation. *Leuconostoc mesenteroids* is the most commonly encountered bacterium followed by , *Streptococcus faecalis* and *Pediococcus cerevisiae*. The yeast flora generally involved in the fermentation include, *Saccharomyces cerevisiae, Deharyomyces hansenil, Hensenula anomala* and *Trichosporon pullulans*.

Materials required: Rice batter, Water blank, petriplates, pipettes and MYGP agar

Composition

Malt extract	0.3g
Yeast extract	0.3g
Glucose	1.0g
Mycopeptone	0.5g
Agar	1.5g
Distilled water	100ml
p^H	5.5

Antimold inhibitor sodium propionate 250 μg/ml

Antibacterial inhibitor streptomycin 100μg/ml

Procedure: 10g of the sample is dissolved in 90 ml water blank. Prepare 10^{-1} to 10^{-4} dilution and plate out dilution of 10^{-3} and 10^{-4} using MYGP agar. Incubate the plates at room temperature and count the yeast colonies.

Result: ....

Ex.No.10. PRESUMPTIVE TEST FOR DETECTING COLIFORM BACTERIA IN FOODS

Introduction: The presence of coliform bacteria in food is an indication of the food contaminated or polluted with the faecal matter. The *Escherichia coli* is a bacterium of the intestinal origin of the human being. The presence of this organism in drinking water shows the chances of the human pathogenic organisms also.

This experiment lies in the principle that the coliform bacteria produce gas in lactose broth within 24 hrs, which constitute the positive presumptive test. The production of gas after 24 hours, but before 48 hours, constitute the doubtful presumptive test.

Materials required: Lactose broth, food sample, Test tubes, Durham's tubes, Sterile pipettes and Sterile water.

Composition of lactose broth

Lactose	:	5 g
Peptone	:	5 g
Yeast extract	:	3 g
Distilled water	:	1000 ml
pH	:	7

Procedure

Prepare the lactose broth and distribute in 10 ml quantities in test tubes. Fill the Durham's tube with lactose broth and carefully push them into the test tubes containing Lactose broth in inverted position without allowing any air bubble. Sterilize the test tube containing a lactose broth at 100°C for 3 successive days. Prepare the serial dilution of the food sample up to 1/1000 dilution. Transfer 1 ml of the dilution 1/10, 1/100 and 1/1000 separately each into 2 test tubes containing lactose broth. Maintain 2 test tubes of lactose broth without adding sample to have as control. Incubate all the test tubes at 37°C for 48 hours. The production of gas is observed by replacement of broth in the Durham's tube. The production of gas within 24 hours is considered as positive for the presumptive test. The confirmed test and completed test are carried out, to further confirm the presence of *E. coli*.

Presumptive test Procedure

1. Inoculation: a) Inoculum 0.1, 1.0 and 10 ml of food sample

   b) Medium lactose broth, fermentation tubes

2. Incubation - 24-48 h at 35°C
3. Results - Gas$^-$: Negative presumptive test
   Gas$^+$: Positive presumptive test

Confirmed test Procedure

1. Inoculation: a) Inoculum from tubes showing gas
   b) Medium brilliant green lactose bile broth, fermentation tubes
2. Incubation 48 h at 35°C
3. Results - Gas$^-$: Negative confirmed test
   Gas$^+$: Positive confirmed test

Completed Test Procedure

1. Inoculation: a) Inoculum from tubes showing gas
   b) Eosine methylene blue agar
2. Incubation - 24 h at 35°C
3. Results - - : Negative colonies
   + : Nucleated purple typical colonies

a. Inoculation: a) Inoculum from typical colifrom colonies
   b) Medium lactose broth with fermentation tubes
b. Incubation - 24-48 h at 35°C
c. Results - No gas from lactose broth : Negative completed test. Gas from lactose broth, gram negative non-spore forming rods : Positive completed test.

Observations :

Ex.No.11. ENUMERATION OF COLIFORMS IN FOODS BY MPN METHOD

Coliforms are group of intestinal bacteria that are aerobic or facultative anaerobic, Gram negative, non-spore forming rods, ferment lactose with gas formation. Coliforms are indicator organisms, closely associated with the presence of pathogenic organisms but not necessarily pathogenic themselves. *Escherichia coli* is a bacterium found in lower intestine of warm blooded animals. Most *E.coli* strains are harmless, but some, such as serotype 0157:H7, can cause serious food poisoning in humans.

Materials required : food sample, buffered peptone water, Test tubes, Durham's tubes, and Lauryl sulphate tryptone broth.

Procedure: Food homogenate is prepared by blending 25g food sample with 225 ml buffered peptone water in a mixer at 1500 -2000 rpm. This 10^{-1} food homogenate is further serially diluted with peptone water to 10^{-2} and 10^{-3}.

Prepare the Lauryl sulphate tryptone broth and distribute in 10 ml quantities in test tubes. Fill the Durham's tube with Lauryl sulphate tryptone broth and carefully push them into the test tubes containing Lauryl sulphate tryptone broth in inverted position without allowing any air bubble. Sterilize the test tube containing LST broth at 100°C for 3 successive days.

One ml each of the above three dilutions are transferred separately to three sets of three Lauryl sulphate tryptone tubes. Incubate all the test tubes at 37°C for 48 hours. The production of gas is observed by replacement of broth in the Durham's tube. Record gas positive tubes and calculate from MPN table.

Result: ....

COLI FORM MPN TABLE

POSITIVE TUBES				POSITIVE TUBES				POSITIVE TUBES				POSITIVE TUBES			
0.1	0.01	0.001	MPN	0.1	0.01	0.001	MPN	0.1	0.01	0.001	MPN	0.1	0.01	0.001	MPN
0	0	0	<3	1	0	0	3.6	2	0	0	9.1	3	0	0	23
0	0	1	3	1	0	1	7.2	2	0	1	14	3	0	1	39
0	0	2	6	1	0	2	11	2	0	2	20	3	0	2	64
0	0	3	9	1	0	3	15	2	0	3	26	3	0	3	95
0	1	0	3	1	1	0	7.3	2	1	0	15	3	1	0	43
0	1	1	6.1	1	1	1	11	2	1	1	20	3	1	1	75
0	1	2	9.2	1	1	2	15	2	1	2	27	3	1	2	120
0	1	3	12	1	1	3	19	2	1	3	34	3	1	3	160
0	2	0	6.2	1	2	0	11	2	2	0	21	3	2	0	93
0	2	1	9.3	1	2	1	15	2	2	1	28	3	2	1	150
0	2	2	12	1	2	2	20	2	2	2	35	3	2	2	210
0	2	3	16	1	2	3	24	2	2	3	42	3	2	3	290
0	3	0	9.4	1	3	0	16	2	3	0	29	3	3	0	240
0	3	1	13	1	3	1	20	2	3	1	36	3	3	1	460
0	3	2	16	1	3	2	24	2	3	2	44	3	3	2	1100
0	3	3	19	1	3	3	29	2	3	3	53	3	3	3	>1100

Ex.No.12. RESAZURIN TEST

Introduction : The resazurin test is a measure of the bacteria in a sample of milk and is based on the effect of bacteria on the oxygen content of milk. As bacteria grow in the milk, consume dissolved oxygen and lower the oxidation - reduction (O-R) potential of milk. More bacteria present in the milk, the faster the reduction of the dye. The two large group of bacteria in the milk, the lactic acid bacteria (*Streptococcus lactis)* and coliforms (*E.coli)* are strong reducers. Resazurin changes colour gradually during reduction from purple to lavender, pink and white (hydroresorufin)

Materials required: Raw milk, pasteurized milk, screw capped vials, pipettes and resazurin solution (11mg resazurin in 20 ml water in an amber coloured bottle)

Procedure: Two milk samples, raw milk and pasteurized milk, are provided. Measure 10 ml of each milk with a pipette and transfer into two screw capped vials containing 1ml resazurin dye solution . Do not mix the dye with the milk. Maintain a control tube without dye. Put the vials in a water bath carefully regulated to 37°C with a thermometer. Then at 37°C mix the contents by inversion and replace. After one hour incubation at 37°C , grade the milk samples in accordance with colour changes given below:

Grade -1	Excellent	- no change from original purple
Grade-2	Good	- Lavender
Grade-3	Average	- Pink
Grade-4	Bad	- white

RESULT

Ex.No.13. METHYLENE BLUE REDUCTION TEST

INTRODUCTION: This test (MBRT) is recommended to determine the microbiological quality of milk. It is based on the effect of bacterial population on the oxygen content of milk. If the sample contains a large population of actively metabolizing bacteria, the populations utilize the oxygen for respiration and the concentration of dissolved oxygen in milk will be low. As the bacteria further grow in the milk, consume the dissolved oxygen and lower the oxidation- reduction potential of milk. When the oxidation- reduction potential is lowered –under anaerobic condition, methylene blue dye in milk will change in to a colourless one, namely methylene white. The time taken for colour change indicates the quality of milk.

Materials required : Raw milk, pasteurized milk, sterile test tubes, pipettes, methylene blue 0.6 per cent, test tube stand.

Procedure: Transfer 15 ml quantity of raw milk or pasteurized milk in to a test tube and add two drops of methylene blue indicator in each tube. Shake the tubes well to mix the dye. Maintain a control tube without the dye. Record the changes in colour from blue to white at periodic intervals up to 360 minutes.

Milk quality Classification is based on time taken for colour change from blue to white.

S.No.	GRADE	TIME TAKEN FOR COLOUR CHANGE	APPROXIMATE POPULATION
1.	VERY POOR	LESS THAN 30 MINUTES	> 20 MILLION
2.	POOR	20 – 120 MINUTES	4 – 20 MILLION
3.	AVERAGE	120 – 240 MINUTES	0.5 – 4 MILLION
4.	GOOD	ABOVE 240 MINUTES	< 0.5 MILLION

RESULT : The time taken for colour change in the milk sample (A) is __ minutes. As per the rating given above the sample A is graded under _______.

Ex.No.14. ROLE OF YEASTS IN BREAD MAKING

Leavened breads are made by adding baker's yeast to dough made from flour, water and salts. Yeasts ferment sugars in dough to alcohol and carbon dioxide. The gas bubbles become trapped in the dough and expand in the heat during baking giving the bread its fluffy texture.

Material required : Baker's yeast, wheat flour, salt, sugar, bread tin, (10x20x10 cm), baking tray and oven

PROCEDURE : Wheat flour (750 g) with a dessert spoonful of salt and sugar is mixed well in a mixing bowl. Yeast cream (starter culture -inoculum) is prepared by mixing 10 g yeast dried powder with little sugar in water and incubated for 15 minutes. During this time yeast begin to breakdown the sugar and thick creamy foam appear. The yeast cream is poured into flour, stirred gently and water (400 ml) mixed well to make smooth spongy dough. The dough is kneaded by pressing with hands for 10 minutes and leaves the dough for an hour. In this time yeast produce carbon dioxide and the volume of the dough is doubled. The dough is packed in bread tins (10 cm height x 10 cm wide x 20 cm length). The dough in the bread tin is as long and as wide as the tin but about half the height. The bread tins are packed in baking tray, during the time the dough rise to fill the tin. The baking trays are transferred to oven set at 240°C and baked for 30 minutes.

Observations : ....

Ex.No.15. PRODUCTION OF SINGLE CELL PROTEIN (SCP)

Single cell protein is defined as the dried cells of microorganisms such as algae, bacteria, actinomycetes and fungi that are grown in large scale culture systems and used as food or feed. The large scale production of microorganisms as a direct source of microbial protein has been well realized. SCP can be used to replace totally or partially the valuable amount of conventional vegetable and animal protein feed. Using microorganisms as SCP has many advantageous. A variety of substrates are used for SCP production. The materials which contain sugars/starch/lignocelluloses can be used as substrates for SCP production. Production of yeast biomass as SCP is aimed in this experiment.

Materials required: Good strain of yeast; Molasses/Sago industry waste, culture medium, container, centrifuge.

Procedure : Grow a good strain of yeast in MYGP broth in a conical flask for about 3-4 days and use it as starter culture. Prepare the substrate in liquid form by mixing the sago industry waste with water or obtain molasses from sugar industry and collect in a container. Inoculate yeast cell suspension @ 10 ml/l of substrate to the container and incubate it till the density of yeast cells reaches to 1.1 g/ml with intermittent shaking at 30-35°C. Harvest the cells by decantation, centrifugation and drying treatment methods. Collect the dried cells and supplement with the animal feed in required proportion. *Spirulina* is a blue green alga used as a potential source of SCP. It grows in waste waters like rice mill effluent and produces SCP photosynthetically as well as recycles wastes water.

The biomass yield coefficient is the efficiency of conversion of substrate to biomass is calculated as

$$\text{Biomass yield} = \frac{\text{Dry weight of biomass produced}}{\text{Weight of substrate used}}$$

Observations : ....

GLOSSARY

Acetic acid bacteria, *Acetobacter aceti, Glucanobacter* sp oxidize ethanol to acetic acid, useful in vinegar manufacture but are undesirable in alcoholic beverages.

Aflatoxins are a complex of mycotoxins produced by species of the soil fungal saprophyte *Aspergillus.* Aflatoxins are a closely related group of secondary metabolites produced by *Aspergillus flavus* and its related species in seeds and food products.

Amylases are hydrolytic enzymes that convert starch into sugars, syrups and dextrins. Fungal amylases for commercial purposes are derived from *Aspergillus oryza* in solid substrate culture with wheat bran as substrate.

Baker's yeast, *Saccharomyces cerevisiae* is used to leaven bread throughout the world. The scientific name *Saccharomyces,* refers to saccharo meaning sugar and **myces** meaning fungus. The species name *cerevisiae,* is derived from the name Ceres, the Roman goddess of agriculture.

Biological acidity is the acidity in fermented milk, pickles, the acidity is due to the action (fermentation) of certain acid producing microbes.

Biological structure is the protective covering in plant and animal foods. Structures such as shell of nuts, rind or skin- the outer covering of fruits and vegetables, shell on eggs, prevent microbial entry and spoilage.

Blanching is a hot water or steam treatment of immersing raw food (vegetables/ fruits) at 63 to 72 °C for 2 to 3 minutes before canning to inactivate endogenous enzymes and reduce microbial load .

Bottom yeast, *Saccharomyces carlsbergensis,* sinks to the bottom of the liquid in beer brewing and ferments best at cool temperatures ranging from 3° to 10° C (38° to 50° F).

Botulism : It is a very serious type of food intoxication disease caused by *Clostridium botulinum.* The disease derives its name from the Latin word Botulus, which means sausage. The major source of botulism is improperly processed home canned vegetable sausage of low acid content.

Breather: This type of spoilage of cans may be due to improper packing or microbial action. The ends of the containers are pushed inside and it is likely to take atmospheric air inside, if holes are found.

Brewing is the production of malt beverages by yeast fermentation. Four basic ingredients are used to brew beer: grain, hops, yeast, and water.

Buttermilk is not high in fat as the name suggest. Actually, the name refers to the fact that buttermilk was once the watery end-product of butter making. Modern buttermilk is made from low-fat or skim milk and has less than 2 percent fat. Its correct name is "cultured low-fat milk" or "cultured nonfat milk."

Canning is a method of preserving food from spoilage by storing it in containers that are hermetically sealed and then sterilized by heat. Canning eliminates all microorganisms that can spoil the food and those that are harmful to humans by exposing the food to high temperature (above100°C) for few minutes or seconds to retain their natural flavour.

Cellulases are microbial enzymes industrially produced from *Trichoderma viridae* and *Aspergillus niger*. Cellulases are used to improve palatability of low quality vegetables. They increase the flavour of mushrooms, promote extraction of natural products and alter the texture of foods.

Cheese is an acidic fermented dairy product made from curd, obtained by the coagulation of milk casein with an enzyme rennin or lactic acid. The curd is further processed with heat, pressure, salt and ripening with selected microorganism to obtain cheese.

Ciguatera poisoning: It is contracted by man upon ingestion of numerous species of shore fish such as barracudas, groupers, sea basses, jacks and eels. They acquire poison as a consequence of feeding on smaller herbivorous fish, which in turn feed on toxic algae (like *Lyngbya majuscule*).

Coliforms are group of intestinal bacteria that are aerobic or facultative anaerobic, gram negative, non-spore forming rods, ferment lactose with gas formation. Coliforms are indicator organisms, closely associated with the presence of pathogenic organisms but not necessarily pathogenic themselves.

Controlled Atmosphere Storage: Fruits and vegetables are stored in sealed warehouses where temperature and humidity are closely controlled. Most importantly, the composition of the atmosphere is controlled by the introduction of carbon dioxide or ozone to minimize spoilage.

Cool storage is the bulk storage of fruits and vegetables in cool environment created by air cooler or water spray. Banana and tomato are stored at 13°C to slow down natural ripening.

Cottage cheese is a un-ripened soft cheese, generally coagulated with lactic acid rather than rennin. The curd is not pressed, aged or ripened.

Cultured butter milk is fresh skimmed pasteurized milk that has been fermented by lactic culture and aroma organisms. The product obtained is a drinkable viscous fluid containing characteristic pleasing aroma and flavour. The bacteria used in fermentation are selected strains of *Streptococcus cremoris* in combination with *Leuconostoc mesenteroides* subsp *cremoris.*

12 D Concept is the heat process required to reduce 10^{12} spores to 1 spore per ml in processed food. As a measure of food safety, a large initial population of the food poisoning bacterium *Clostridium botulinum* is assumed to be present.

D value (Decimal reduction time) is defined as the time of heating at a temperature to cause a 90 per cent reduction in the count of viable cells or spores in a substrate (food). D value gives the resistance of an organism to a specific temperature.

Dust bacteria are commonly found in air and can endure dryness e.g. *Bacillus, Micrococcus, Streptococcus* and *Sarcina* species.

Endotoxin is bound to cell wall and liberated only when the cell producing it disintegrates. The toxic principle is carbohydrate, lipopolysaccharide (LPS) or lipid. It is generally produced by Gram negative bacteria and toxicity is low. e.g. Salmonella toxin.

Enterotoxin causes gastroenterititis, the inflammation of the lining of intestinal tract. The principal symptoms are nausea, vomiting, abdominal pain (cramps) and diarrhea. e.g. toxins produced by *Salmonella, Vibrio, Shigella, E. coli, Staphylococcus, Bacillus cereus.*

Ergotism is the name for severe pathological syndromes affecting humans or animals that have ingested ergot alkaloid-containing plant material, such as ergot-contaminated grains.

ETEC (Entero toxigenic *E.coli*) strains are the primary cause of infant diarrhea and *travellers' diarrhea.*

Exotoxin is secreted by the bacteria into surrounding medium. The toxin, denatured by heat, is a type of protein produced by Gram positive / negative bacteria. The toxicity is great. e.g. Botulin toxin.

Fermented foods are defined as food that has been subjected to the action of selected microorganisms by which a biochemically and organoleptically modified substrate is produced.

Flat sour spoilage : The ends of canned food remain flat, have a normal concavity, during souring by flat sour bacteria. The spoilage occurs in low acid foods (pea or corn), is caused by *Bacillus* spp.

Food intoxication refers to food borne illness caused by the ingestion of a microbial exotoxin preformed in the ingested food. e.g. Botulism.

Food infection refers to food borne illness caused by the entrance of viable, usually multiplying pathogenic bacteria into the body through the ingestion of contaminated food. e.g. Samonellosis.

Food irradiation is a process in which food products are exposed to a controlled amount of radiant energy to kill harmful microorganisms, insects and parasites and prevent food spoilage.

Food packaging: Package contains food and makes it easier to handle, and protects it from environmental conditions, such as temperature extremes, during transport. It locks out microorganisms and chemicals that could contaminate the food, and helps prevent physical and chemical changes and maintain the nutritional qualities of food.

Food poisoning is an acute illness resulting from the ingestion of injurious agents such as chemical, heavy metals, parasites, microbial toxin, wild species of mushroom in food. Food poisoning is commonly caused by certain bacteria or their toxins.

Food preservation: Refers specifically to the processing techniques that are used to keep food from spoiling and to permit its future consumption. Preservation is a process of inhibiting, retarding and arresting the growth of spoilage microorganisms.

Food Safety is related to the presence of food borne hazards in food at the point of consumption. One weak link can result in unsafe food that is dangerous to health. As food safety hazards can occur in the food chain at any stage, adequate control throughout the supply chain is essential.

Food spoilage may be defined as any change that renders food unfit for human consumption. These changes may be caused by various factors, including contamination by microorganisms, infestation by insects, or degradation by endogenous enzymes (those present naturally in the food).

Freeze-drying or **lyophilization**, a preservation method in which moisture in the food is frozen and then sublimed under vacuum - a process in which ice changes from a solid directly to a vapour without first becoming a liquid.

Freezing preserves food at low temperature (below 0°C) by preventing microorganisms from multiplying and retarding the action of food enzymes. Freezing usually results in a considerable reduction in number of viable microorganisms due to its lethal and sub-lethal effects.

Fumonisins are mycotoxins produced throughout the world by *Fusarium moniliforme* and related species when they grow in maize.

Gas forming bacteria include species of the genera, *Leconostoc, Lactobacillus and Propionibacterium* which produce carbon dioxide. Species of *Escherichia, Enterobacter, Proteus, Bacillus, and Clostridium* produce both carbon dioxide and hydrogen.

Genetic engineering refers to any deliberate alteration of an organism's DNA. Genes have been cut and pasted from one species to another, yielding, for example, disease-resistant squash, frost-resistant potatoes, golden rice and slow ripening tomatoes.

Genetically modified organism (GMO) is a microbe or more complex plant or animal in which DNA has been altered for a particular purpose such as manufacturing proteins , making improvement to the organism , research into the nature of genes and biological process, correcting genetic defects.

Golden rice: It is a transgenic rice crop, genetically engineered to produce β-carotene, pro vitamin A. developed by Dr. Ingo Potrykus, Swiss Genetist.

Good manufacturing practices (GMP) is defined as a set current scientifically sound methods/practices/principles, that are implemented and documented during food production to ensure consistent manufacture of safe, pure and potent products.

HACCP is a food safety system based on a set of safety regulations. Hazard **analysis** is aimed at identifying food hazards, biological, physical and chemical contaminants and maintains safety of foods. Critical control point is a point or step or procedure where control can be applied and hazard can be prevented , eliminated or reduced.

Halophilic bacteria, salt loving organisms, require certain minimal concentrations of dissolved sodium chloride for their growth and survive at higher salt concentrations to different extents depending on their halotolerant ability, include species the genera *Bacillus, Vibrio, Moraxella, Halobacterium, Corynebacterium.*

Heterolactic bacteria – *Leuconostoc* and *Weissella,* follow 6 phosphogluconate pathway to transform a glucose molecule into lactate, ethanol, and carbon dioxide.

High pressure processing (HPP) or High hydrostatic pressure (HHP) or ultra high pressure processing (UHP) is a non-thermal food preservation technique that subject liquid and solid foods with or without packaging to pressures between 100 and 800 MPa.

Homolactic-bacteria such as *Lactococcus* and *Streptococcus* follow glycolytic pathway to yield two lactates from one glucose molecule.

Hops are obtained from dried flowers of a plant, *Humulus lupulus.* The boiling of wort (in beer brewing) with hops extracts tannins, essential oils, bitter acids and resins.

Idlies in the form of spongy pancakes are consumed as a breakfast food in south India since, 1100 A.D. They are auto-fermented foods of acidified and leavened type. These are prepared by steaming the fermented batter consisting of rice and black gram grains and served with spicy hot side dishes (sambar or chutney).

Immobilization of enzyme has been defined as the "imprisonment of an enzyme in a distinct phase that allows exchange with, but is separated from, the bulk phase in which substrate, effector or inhibitor molecules are dispersed and monitored".

Inhibitory substances in food are antimicrobial substances present in food that prevent the microbial growth and multiplication. They may be, originally present in food, or developed by the growth of microbes or added purposely or accidentally.

Invertase is widely used for hydrolysis of sucrose. The hydrolysis (inversion) of sucrose to glucose and fructose facilitate more sweetness, enhancement of fruit flavour , high stability at low pH and is less likely to crystallize in products. The yeast invertase produced by *Saccharomyces cerevisiae* is used in confectionery industry and to make syrup of invert sugars.

Kefir grains are whitish or yellowish gelatinous irregular granules of varying size and largely made up of a polysaccharide called kefiran. The granules are insoluble in water and swell up when soaked in water to form a jelly. Within the folds of granules the mixture of at least six functionally different groups of bacteria (*Lactobacillus, streptococcus, Micrococcus, Bacillus*) and yeasts (*Saccharomyces kefir, Torula kefir, Candida*) are found.

Kefir is a acid-alcoholic fermented milk with high therapeutic value, popular in Russia and Central Asia. Kefir is made from whole milk of goats, sheep or cows held at 85° C for 30 minutes. The heated milk is cooled to 22° C, inoculated with kefir grains and incubated over night at 22 ° C. The resultant smooth curd has 1.0 per cent lactic acid, 0.5 percent alcohol and 0.05 percent carbon dioxide.

Kilogray (KGy), the unit of absorbed dose of radiation by a material. One gray is equal to the absorption of one joule of energy by one kilogram of food.

Koji mold refers to filamentous molds grown or cultured on solid plant materials in food processing as hydrolytic enzyme producers. It includes Aspergillus molds *Aspergillus oryzae, A.sojae, A. tamari, A. awamori, A. usami* and molds in genera *Rhizopus* and *Mucor.*

Koumiss is a acid-alcoholic fermented milk with high therapeutic value, popular in Russia and Central Asia. It is traditionally made from fresh mares milk warmed to 28 °C. The milk is then inoculated with 30% of a starter consisting lactic acid bacteria (*Lactobacillus delbrueckii* subsp *bulgaricus*) and yeasts (*Torula kefir*).

Lactase or β galactosidase enzyme that converts lactose in to glucose and galactose, is produced by *Aspergillus niger, A.oryzae , Kluyveromyces fragilis*

Lactobacillus amylophilus are used in the production infant milk and solubility of milk products.

Lactic acid bacteria ferment sugars to lactic acid and include the species of *Leuconostoc, Lactobacillus, Streptococcus* and *Pediococcus.* They are normally present in milk, and are also used as starter cultures in the production of fermented dairy products.

Lantibiotics are bacteriocins produced by LAB, which have a long history of use in fermentation and meat and milk preservation, are the best characterized of this group.

Lipolysis is the hydrolysis of milk fat by lipase resulting in accumulation of free fatty acids. Lipolysis of milk fat and the subsequent production of off-flavours in dairy products is a potentially important food spoilage problem in cream and butter.

Lipolytic bacteria are a heterogenous group of bacteria, which produce lipases. Many of the aerobic proteolytic bacteria are also lipolytic. Most species belong to the genera, *Pseudomonas, Alcaligenes, Staphylococcus*, *Serratia* and *Micrococcus.*

Miso is a semisolid salty food prepared using koji mold. Rice miso is produced using rice and cooked soya beans, Barley miso from barley and cooked soya beans and soya bean miso from cooked soya beans.

Modified atmosphere packaging (MAP): In this method the air surrounding the food product is replaced by a different mixture of gases, where the proportion of each component is fixed when the mixture is introduced, but no further control is exercised during storage. Oxygen, Nitrogen and Carbon dioxide are the three major gases commercially used in MAP.

Mushrooms are most ancient, best known and greatly admired fleshy macrofungi of the order *Agaricales* in the class Basidiomycetes but also of some other groups. Popularly, the term mushroom is used to identify the edible sporophores. They are delicacies, much sought after and greatly valued foods.

Mycotoxins are non-volatile, toxic secondary metabolites of low molecular weight produced by certain saprophytic and pathogenic fungi on food materials under both pre and post harvest conditions of tropical and subtropical regions.

Natural acidity is the acidity in fruits, the acidity is due to organic acids of plant origin (citric acid in lime).

Neurotoxin is a exotoxin produced by *Clostridium botulinum* affects muscles of vision, swallowing and respiration. The toxin blocks acetylcholine at neuromuscular junctions, halts neuronal activity and muscles are paralyzed.

Niacin is the best known bacteriocin most frequently studied anti-microbial low molecular weight protein produced by *Lactococcus lactis* subsp. *lactics*. It is widely used in various foods such as dairy products, canned foods, meat products, etc. Niacin inhibits a wide range of Gram positive bacteria, particularly those that produce spores.

Non-perishable or stable foods : These foods , which do not spoil unless handled carelessly, include such products as sugar, flour and dry beans.

Ochratoxins are mycotoxins produced by *Aspergillus* and *Penicillium* strains, found on cereals, coffee and bread, as well as on all kinds of food commodities of animal origin.

OR potential (oxidation-reduction potential) of a substrate may be defined as the ease with which the substrate gains or losses electrons. When it losses electrons, it is oxidized and while it gains electrons it is said to be reduced.

Osmophilic yeasts have high tolerance for high solute concentrations. They concentrate polyols as osmoregulators and enzyme protectors. Osmophilic yeasts, *Zygosaccharomyces rouxii, Z. bisporus, Z. baili* cause spoilage of sugar syrups, fruit concentrates and chocolates.

Oyster mushroom- *Pleurotus* spp are the easiest, fastest and cheapest to grow, require less preparation time and production technology. With more than 100% biological efficiency, coupled with distinct oyster like flavour, aroma and excellent drying and preservation qualities it is assured a unique status as a delicacy. This mushroom grows in bracket-like clusters.

Paddy straw mushroom- *Volvariella volvacea*- It is mainly grown in paddy straw, hence the name Paddy straw mushroom. These fungi derive their carbonaceous food by decomposing lignin, cellulose and hemicellulose present in agro-wastes with the help of extracellular enzymes secreted by the mycelium.

Pasteurization is defined as a process of heating a liquid food to a controlled temperature below its boiling point for a specific time, to enhance the keeping quality and destroy harmful microorganisms without changing the composition, flavour, taste or nutritive value of the liquid food (Milk, Wine).

Pectinases/Pectic enzymes are produced by species of *Aspergillus , Botrytis, Penicillium, and Rhizopus*. They are added to fruit pulp to hydrolyze the pectin, so that the juice becomes clear.

Pectinolytic bacteria produce pectinases responsible for softening the plant tissue or loss of gelling power in various plant foods. Species belong to the genera, *Bacillus, Achromobacter, Aeromonas, Arthrobacter, Flavobacterium* are pectinolytic.

Perishable foods: This group includes most of our important daily foods that spoil readily unless special preservation methods are used. e.g. Fruits and vegetables, meat, fish, poultry, eggs and milk.

Phytases are enzymes that hydrolyze phytic acid in to myo-inositol and inorganic phosphates, produced by *Aerobacter aerogenes* and *Bacillus subtilis.* Phytic acid in food ingredients, wheat flour, rice flour, gram flour and animal feed supplements are dephytized by phytase.

Pigmented bacteria produce colour during their growth in foods. e.g. *Flavobacterium* (yellow to orange), *Serratia* (red), *Halococcus* (pink, red and orange). *Lactobacillus plantarum* produces rust colour pigment discolouring cheese.

Probiotic, which literally means "for life," is employed to describe these health-promoting bacteria. It is defined as "live microorganisms which when administrated in adequate amounts confer a health benefit on the host by improving the intestinal microbial balance".

Proteases are hydrolytic enzymes industrially produced by species of *Aspergillus oryzae, Mucor and Rhizopus.* These protein splitting enzymes from microorganisms are widely used in protein industries, where the protein containing materials are processed into foods and feeds. This imparts improved textural, nutritional and sensory properties to the products and in addition making maximum utilization of proteins.

Proteolytic bacteria include a heterogenous group of bacteria, which produce extracellular proteases. Most species belong to the genera, *Clostridium, Bacillus, Pseudomonas* and *Proteus.*

Pruteen process is a process developed by ICI utilizing *Methylophilus methlyotrophus - SCP* grown continuously on a sterile medium containing methanol, inorganic salts and ammonia.

Psychrophilic bacteria are cold loving, able to survive and grow at refrigeration temperature. As they can grow relatively rapidly at low temperature ranges around 0°C they spoil non-sterile refrigerated foods such as meat, poultry, egg and other foods. e.g. *Pseudomonas, Achromobacter, Alcaligenes, Chromobacterium* and *Flavobacterium.*

Putrefaction is the anaerobic decomposition of proteins, polypeptides, simpler peptides or amino acids that may result in the production of foul smelling obnoxious odours due to sulphur containing products such as hydrogen sulphide, mercaptans, plus ammonia, amines, indole and fatty acids.

Radappertzation is radiation sterilization or commercial sterility of canned foods to eliminate endospores and exotoxin of *Clostridium botulinum.*

Radicidation is equivalent to radiation bactericidation that reduces the non-spore forming pathogens other than viruses and present no health hazard.

It is effective on pre-packaged food products (raw poultry, spices and condiments) and eliminates the possibilities of cross contamination.

Radurization or radiation pasteurization refers to the application of small amounts of beta or gamma rays to foods to increase their storage time. Radurization is equivalent to pasteurization to enhance the keeping quality and shelf life of fresh meat, poultry, sea food, fruits, vegetables and cereals by causing reduction in spoilage microbes by radiation.

Red wines are prepared by fermentation of grape juice along with the grape skins. Red pigments called anthocyanins and other compounds in the grape skins are extracted during the fermentation process to impart the characteristic red color of the wine as well as other features.

Rennets are proteolytic enzymes that are used for the coagulation of milk in cheese production. The most common rennet enzyme in commercial use throughout the world is calf rennet, an enzyme extract obtained from the fourth stomach (abomasums) of calves.

Ropiness/ Sliminess: Ropy fermentation of milk and cream is a spoilage brought about by the growth of bacteria leading to change in consistency of the product that forms threads or viscous masses when poured. The causative organisms are Gram negative rods- *Alcaligenes viscolactis, Achromobacter, Flavobacterium.* The gums (polysaccharides) derived from lactose are responsible for ropiness caused by coliforms.

Saccharolytic bacteria: Hydrolyze disaccharides and polysaccharides to simple sugars. *Clostridium butyricum* and *Bacillus subtilis* are examples of saccharolytic bacteria, which are also amylolytic, produce amylase to hydrolyze starch.

Sake : Rice wine is the Japanese national drink prepared by fermenting rice with *Aspergillus oryzae.*

Salmonellosis is a food borne infection that may result following the ingestion of viable cells of *Salmonella typhimurium, S. enteritidis, S.cholerusuis.*

Sauerkraut: (German terms for sour cabbage), is a product of natural lactic acid fermentation of fresh cabbage in the presence of a small amount of salt involving lactic acid bacteria (LAB*), Leuconostoc mesenteroides* and *Lactobacillus plantarum.*

Semi-perishable foods: If these foods are properly handled and stored, they will remain unspoiled for fairly long period. e.g. potato, onion.

Shigellosis or bacterial dysentery is a food borne infection caused by four species of pathogenic *Shigella,* including *S. dysenteriae, S.sonnei, S.boydii* and *S. flexneri*. These organisms can persist in foods for up to a month.

Shiitake mushrooms: *Lentinus edodus* - This is the major edible mushroom grown in Japan. This fungus produces attractive brown-red fruit bodies and sold as both as the fresh and dried product. Shiitakes are grown by inoculating, mushroom spawn into holes drilled into suitable species of hardwood logs.

Single cell oils are the microorganisms, *Fusarium pallidoroseum, F. lycopersici, Aspergillus sydowii* that produce 45 to 50 % oil from simple carbon sources. SCO organisms are potential for major modification in the fatty acid composition of their oils through selection or recombinant techniques.

Single cell proteins (SCP) are the dried cells of microorganisms such as bacteria, fungi, yeasts, algae that are grown in large scale culture systems as proteins for human or animal consumption.

Smoking, is a preservation method, applied to preserve fish, ham, and sausage. The smoke is obtained by burning hickory or a similar wood under low draft. In this case some preservative action is provided by such bactericidal chemicals in the smoke as formaldehyde and creosote, and by the dehydration that occurs in the smokehouse.

Sour cream is a fermented dairy product results from souring by lactic acid producing bacteria of pasteurized cream. Sour cream contains not less than 18 % milk fat. Sour cream is used as lopping on baked potatoes, vegetables, salads and fruits.

Soy sauce : It is a Savory flavoured sauce traditionally prepared by fermentation of soya beans, with the mold *Aspergillus oryzae* which transform bland vegetable protein into meat flavoured sauces.

Spirulina are multicellular, filamentous, non-heterocystous, non-nitrogen fixing cyanobacteria characterized by spiral structure. They have many spirals, appear as green filaments composed of cylindrical cells. They are found in fresh water streams, brackish water, sea, tidal ponds and saline ponds. Spirulina contains around 60% protein, 6% crude lipids, 15% carbohydrates, 0.1 % β carotene and can play an important role in human nutrition as a protein supplement (SCP).

Springer: The container appears bulged in one end of the can or both the ends. The bulging resembles as if the container is dashed against a hard object. This is due to the production of acid as well as gas by the bacteria causing this spoilage of canned food.

Sulphide spoilage: The bluish gray liquid in canned corn food is evident of the spoilage and caused by *Clostridium nigrificans*. It is detected by means of black (FeS) colonies it forms in an iron sulphite agar at 55°C.

Sweet curdling is a defect associated with milk and cream, the dairy product become solid, although no acid has been produced. The defect is due to the production of an extra cellular enzyme, which causes casein to precipitate in the form of small specks of curd. The causative organisms are aerobic spore formers- *Bacillus cereus, B. subtilis, B.mycoides.*

Swiss cheese is a hard type cheese, characterized by large holes or eyes and sweet nutty flavour, obtained through the activities of *Propionibacterium shermani.*

TA spoilage (Thermophilic anaerobic spoilage): This spoilage is caused by thermophilic anaerobic bacteria like *Clostridium thermosaccharolyticum,* which form acid and gas (carbon dioxide) in low acid and medium acid canned foods.

Tempeh is a compact moldy white solid soya bean product., occupies a prominent place among popular fermented foods in South Asian region. The inoculation with *Rhizopus oligosporus* spores and allowing to ferment semi-aerobically for about 36 hours to get a compact moldy white solid product. Beside, tempeh contains vitamin B_{12}, enhanced isoflavanoids as anti-oxidants and more available iron.

Thermal death point (TDP) is defined as the temperature necessary to kill all the organisms in a fixed time, usually ten minutes.

Thermal death time (TDT) is defined as the time required at a certain temperature to kill a stated number of microorganisms under specified conditions. The heat resistance of microorganisms is expressed in their thermal death time.

Thermoduric bacteria survive heat treatment such as pasteurization. e.g. *Micrococcus, Staphylococcus, Microbacterium.*

Thermophilic bacteria are resistant to high temperature (above 60° C) . The thermophilic bacteria like *Bacillus stearothermophilus* and *Clostridium thermosaccharolyticum* cause spoilage of low acid canned foods.

Top yeast, *Saccharomyces cerevisiae*, that floats on top of the liquid in beer brewing as it ferments the grain sugars and prefers warm temperatures ranging from 16° to 22°C (60° to 72°F).

Total Quality Management (TQM) system, is a combination, coordination and compliance of all safety and quality control measures with quality assurance programmes are to be followed by food processing industries.

Toxigenic types : There are seven antigenically distinct toxigenic types of *Clostridium botulinum* namely, A,B,C.D.E.F and G. Human botulism is always result from ingestion of toxins from types : A,B,E,F and G. The C and D toxins cause diseases in animals.

Trichothecenes are mycotoxins produced mostly by *Fusarium graminarium,* although other genera, *Trichoderma, Trichothecium, Myrothecium* and *Stachybotrys* are also known to produce these compounds.

Ultraviolet radiation treatment: It is one the non-thermal preservation method used for fruit and vegetable juices. UV-C(200-280 nm) is considered germicidal range since it effectively inactivates bacteria and viruses. UV light kill or interrupt bacterial replication due to the formation of thymine dimers in bacterial chromosome.

Vinegar is literally translated as 'sour wine', is one of the oldest products of fermentation used by man. It is made from fermentable carbohydrate source like fruit, vegetables, syrups and wine. Acetic acid is produced by the oxidation of ethyl alcohol by bacteria such as *Acetobacter aceti.*

Water activity, a_W is defined by the ratio of vapour pressure of water in a product (within food) to the vapour pressure of pure water at the same temperature. The a_W for pure water is 1.0, of most fresh foods is 0.99 and of saturated solution of NaCl is 0.75.

Whey, the milky liquid separated from the curds during cheese making.

White wine, which is actually straw to golden-yellow in color, is produced from white grapes, by taking the grape juice only. White grapes are crushed and the juice separated from the skins prior to fermentation.

World Food Day is celebrated every year on 16 October to commemorate the founding of the Food and Agriculture Organization of the United Nations (FAO) in 1945. World Food Day aims to heighten public awareness of the plight of the world's hungry and malnourished and to encourage people worldwide to take action against hunger.

Yersiniosis is a food borne infection with *Yersinia enterocolitica,* Gram negative bacillus, that cause enterocoliticus-gastroeneritis in children, mesenteric adenitis in older children that may mimic appendicitis and systematic disease in adults characterized by bacteremia. It is generally associated with under cooked beef and pork and milk.

Yoghurt, fermented, semi-fluid milk product, is used extensively as a health food. It is prepared from fresh whole or skim milk, boiled and concentrated by evaporation. Fermentation is caused by the addition of mixed culture of lactic acid bacteria, usually *Streptococcus thermophilus* and *Lactobacillus bulgaricus* (sometimes *Lactobacillus acidophilus*) in a ratio of 1:1, at 35 to 45°C.

Z Value is defined as increase in temperature (in Fahrenheit) required to reduce the processing time (D value) by ten times. Z value provides information on the relative resistance of an organism to different destructive temperatures.

FSC
www.fsc.org
MIX
Papier | Fördert gute Waldnutzung
FSC® C083411

Zeitfracht Medien GmbH
Ferdinand-Jühlke-Straße 7
99095 Erfurt, Deutschland
produktsicherheit@kolibri360.de